The Simple Art of SoC Design

Michael Keating

The Simple
Art of SoC Design

Closing the Gap between RTL and ESL

 Springer

Michael Keating
Synopsys
1925 Cowper St.
Palo Alto, CA 94301
USA
e.mike.keating@gmail.com

ISBN 978-1-4419-8585-9   e-ISBN 978-1-4419-8586-6
DOI 10.1007/978-1-4419-8586-6
Springer New York Dordrecht Heidelberg London

Library of Congress Control Number: 2011924222

Printed on acid-free paper

Springer is part of Springer Science+Business Media (www.springer.com)

Disclaimer

Because of the possibility of human or mechanical error, neither the author, Synopsys, Inc., nor any of its affiliates, including but not limited to Springer Science+Business Media, LLC, guarantees the accuracy, adequacy or completeness of any information contained herein and are not responsible for any errors or omissions, or for the results obtained from the use of such information. THERE ARE NO EXPRESS OR IMPLIED WARRANTIES, INCLUDING, BUT NOT LIMITED TO, WARRANTIES OF MERCHANTABILITY OR FITNESS FOR A PARTICULAR PURPOSE relating to this book. In no event shall the author, Synopsys, Inc., or its affiliates be liable for any indirect, special or consequential damages in connection with the information provided herein.

Foreword

A new graduate may think: "SoC design is exciting; I want to design chips for SmartPhones!" But, experienced RTL designers know that the reality of SoC design is more than exciting. It takes blood, sweat and tears to wrestle up to 20 Million lines of Verilog code into a production-ready product. Chip companies apply man-power, the latest tools and very sophisticated methodologies to find and fix the bugs in an SoC before it goes to silicon – bugs that can run into the thousands.

Hardware designers take pride in the fact that, out of necessity, they routinely create higher quality code than software developers. Unlike software that is often fixed after the product is shipped, the hardware must be essentially bug-free before tape-out. As design size and complexity has grown dramatically, it has become much harder for hardware design teams to live up to this promise. The result is often called the "verification crisis." The design community, along with the EDA industry, has responded to this crisis by making significant improvements in verifi-cation technology and methodology. The testbench has become parameterized and object oriented, and the evaluation of simulation results is now automated. This has helped to make verification teams much more productive.

But it is obvious that the verification crisis cannot be solved exclusively on the verification side. It has to be addressed on the design side as well. Why fix bugs in your design, if you can avoid them in the first place? Why create more bugs than necessary by writing too many lines of code? There are several approaches to the problem. The current generation of high-level synthesis tools allows for a drastic reduction in code size and thus reduces the number of bugs a designer will intro-duce. They generate good quality implementations for a wide range of signal processing applications. This is closing the gap from the top. The other approach is to move RTL designers incrementally up to the next level, improving quality while staying within the RTL paradigm that they are comfortable in.

In this book, Mike Keating takes on the design part of the problem from the pragmatic view of an RTL designer. As co-author of the *Reuse Methodology Manual* and the *Low Power Methodology Manual*, he has established a track record of delivering practical design methodology. In this book, based on his extensive experience and research, Mike proposes some very practical, proven methods for writing better RTL, resulting in fewer lines of code and fewer bugs. He calls writing RTL an art, but he also realizes that every artist deserves the best tools –in this case a language that facilitates good design. To this end he suggests how the language (SystemVerilog) could be extended to enable a better, more concise coding style.

Whether you are a college student or an experienced RTL designer, I hope you will be open for change in how hardware design is done. We at Synopsys have supported Mike Keating's work on this book, because we firmly believe that we need to get new concepts in front of RTL designers. We feel that a strong collaboration between designers and the EDA industry is key to designing tomorrow's most advanced SoCs.

VP of Strategic Alliances, Synopsys   Rich Goldman

Preface

On a bleak January night in 1992, I sat hunched over a computer screen and a logic analyzer. It was well past midnight, and I was the only person in the lab – probably the only person in the building. We had just gotten an ASIC back from the ASIC house, and, of course, when we fired it up it didn't work. The other guys had narrowed the problem down somewhat; now it was my turn to try to find the cause.

I had narrowed it down to a particular module which had been written by an engineer we'll call Jeff. Working my way through Jeff's code, trying to find the cause of the bug, I realized that he had not indented his *if-then-else* statements. This made it absolutely impossible for me to figure out what his code was doing. So at 1:00 in the morning, I spent an hour or so carefully indenting his code – and thinking very unkind thoughts about Jeff. Once I had his code carefully laid out, it was trivial to find the problem - it was an *else* that should have been associated with a different *if*. Of course, with such poorly structured code, it is unlikely that Jeff knew exactly what his code did. Otherwise he would have spotted the rather obvious problem himself. Instead, the problem made its way into the silicon. Fortunately, we were able to compensate for with a software change.

In early 2010, I happened to interview a significant number of candidates for an entry-level design position. Most of these candidates were right out of school, but a few of them had a couple of years of experience. To each of these candidates I gave a very simple problem, to set them at their ease before I started asking the hard ones. I drew this on the whiteboard:

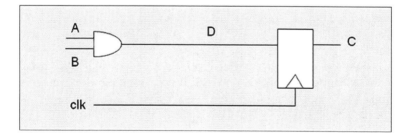

Every single candidate who was right out of school said (pretty much word for word):

"Let us label the output of the AND gate D."

Then they wrote:

```
assign D = A && B;
always @ (posedge clk) begin
   C <= D;
end
```

None of the experienced candidates wrote it this way. And it never occurred to me that anyone in their right mind would. The experienced folks all wrote:

```
always @ (posedge clk) begin
   C <= A && B;
end
```

This answer is simple and straight to the point. Why would you ever add an extra line (and an extra process) to the code? It makes no real difference in this trivial example. But when one is hunched over a computer screen and a logic analyzer in the small hours of the morning, extra lines of code and extra processes can become very expensive. Especially in designs that are tens or hundreds of thousands of lines of RTL code, all written by someone else.

These two events have bracketed twenty years of trying to deal with, and improve, code based design. As a manager and as a researcher, I have spent far more time reading other peoples' RTL than my own. I have watched how engineers struggled when they had to debug someone else's code – often to the point where it was easier to re-write the whole module than find the bug.

In running an IP development team, I learned how critical quality is, and yet how difficult it is to achieve zero-defect code. For years I thought this was a verification problem. I am now convinced that the problem is how we design hardware and how we write RTL code.

Recently I have spent a lot of time looking at high level design and synthesis tools. I think there is some real value in the approaches they take. But I think there are significant opportunities for improvement in how they approach the design problem. Most importantly, I have come to realize that good design and good code do not miraculously emerge from raising abstraction.

Good design and clean code are a fundamental challenge to the human intellect – to make simple the complex, to make clear the obscure, and to add structure to what can look like complete chaos.

This book is an attempt to frame and answer the question of what makes good design and clear code. It presents my conclusions from the last twenty years of struggling with the problems and challenges of designing complex systems – and in particular, the design of SoCs and the IP that goes in them.

It would be impossible to thank all the people who have helped me as I developed (and borrowed, and occasionally stole) the ideas presented in this book. Many of my friends and colleagues - both inside and outside Synopsys – have spend numerous hours talking (and arguing) over these issues.

But I would like to thank specifically the IP development team in Synopsys, including Subramaniam Aravindhan, Steve Peltan, James Feagans, Saleem Mohammad, Matt Meyers, and Qiangwen Wang.

I'd like to thank Tri Nguyen, Aaron Yang, and Shaileshkumar Kumbhani who as interns helped with many experiments whose results have shaped my thinking and the conclusions discussed in this book. I am happy to report that they all now have real jobs as IP developers.

Finally, I'd like to thank Jason Buckley, Badri Gopalan, Arturo Salz, Dongxiang Wu, Johannes Stahl, Craig Gleason, Brad Pierce, and David Flynn for their valuable discussions and feedback on the manuscript.

Contents

Chapter 1
The Third Revolution

The Problem

As semiconductor technology relentlessly pursues the path described by Moore's law, the challenges of SoC design continue to grow dramatically. We are moving from chips with millions of gates to ones with billions of gates. The task of designing such complex systems is becoming extremely difficult – and very expensive. Figure 1-1 shows an estimate of the escalating development cost for a complex SoC as we have moved from 130nm to 22nm technologies.

The explosive growth in the cost of chip design is driven by software and verification. In a very real sense, these two issues are the same: the difficulty in writing, testing and debugging code.

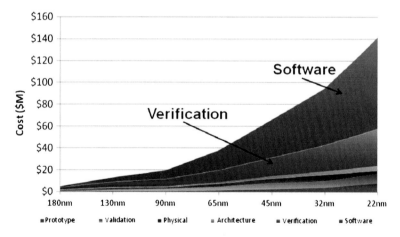

Figure 1-1   Source: International Business Strategies, Inc. (Los Gatos, CA). Used by permission.

M. Keating, *The Simple Art of SoC Design: Closing the Gap between RTL and ESL,*
DOI 10.1007/978-1-4419-8586-6_1, © Synopsys, Inc. 2011

By contrast, the cost of physical design of chips has remained relatively stable over the last few technology generations. Place and route are well-defined, but computationally intensive tasks. The rapid growth in compute power - and the software that takes advantage of it - has kept the overall cost of physical design under control.

But writing and verifying the code for the chip - the RTL and the software that describe the intended function of the chip – are a completely different matter. The languages, tools, and methodology we use for code development have changed little as chips have become larger and more complex. The basic design approach for digital design – how we code Verilog or VHDL to describe complex functionality – has remained largely unchanged in the last fifteen years.

The code development numbers tell the tale:

Studies show that over the life of a project, RTL and software engineers average about 10 to 30 lines of code per day. This includes specifying, writing, testing and debugging of the code [4a][4b][4c]. This number has remained roughly constant for decades. If anything, lines of code per day may be decreasing as chips become larger and more complex. Studies show that code productivity drops as the size of the project increases [5].

The amount of functionality per line of code has also remained roughly constant – at about three to ten gates per line of code of RTL. The lower estimate is typical of control dominated code; the upper estimate is typical of data path dominated code.

Using a cost per engineer at $150k per year and the (upper end) estimate of 30 lines of code per day, we can calculate:

$150k per year/260 work days per year ≈ $600 per day
$600 per day/ 30 lines of code per day = $20 per line of code
$20 per line of code / 10 gates per line of code = $2 per gate

This is an optimistic cost analysis. Studies of commercial software projects suggest an average cost that is slightly higher: about $25 to $33 per line of code [24][25] [26][27]. Our experience with many chip and IP design projects supports the view that these numbers are typical of RTL code development as well.

This cost per gate and cost per line of software code have remained constant from the 180nm design of a few years ago to the leading edge 22nm designs of today. But the amount of software functionality and the number of logic gates in these designs roughly doubles with each generation.

Over the last fifteen years, design reuse has been used extensively to help reduce the cost of chip design, and to help offset the cost of RTL development. But reuse cannot, on its own, solve this problem. To develop differentiated chips, we must develop new hardware and software. And in turn, IP blocks are becoming increasingly large and complex, escalating their cost as well.

The result is that the cost of developing code for complex SoC designs (and the IP that goes into them) is growing exponentially, and swamping other design costs. We have responded to this challenge by adding more and more engineers to each project – and decreasing the number of new design projects.

This trend is not sustainable. We are rapidly reaching a crisis point where we will be limited not by what we can manufacture but by what we can design. It is time to explore how to migrate our design methodology, tools, and languages forward to meet the challenges of designing the highly complex digital systems of tomorrow.

Over the last twenty-five years, there have been two major revolutions in how we do hardware design. The first, starting in 1986, was the move from schematic-based design to RTL and synthesis. The second, starting in 1996, was the adoption of design reuse and IP. We are overdue for the third revolution. This book attempts to describe the initial steps in this revolution. We start with a discussion of the fundamental aspects of good design.

Divide and Conquer

Divide and conquer is the key tool for solving many complex problems. The effective design of complex systems relies on the same principle: the partitioning of the system into appropriately sized components and designing good interfaces between them.

This book explores how to apply this fundamental technique to the design of complex hardware, in particular to SoC design.

The process of designing complex chips is itself a complex system. In the early days, the Integrated Device Manufacturer (IDM) model was common. A single company (such as LSI Logic in the 1980's) could have its own EDA tools, its own fab, and its own design teams. Today, this complex system has been partitioned into separate subsystems.

EDA companies develop the tools to do design and provide them to the entire community. In general, design tools are simply too complex for design houses to develop on their own.

Similarly, fabs have become too expensive for most semiconductor companies to build and maintain. Instead, independent fab houses such as TSMC, UMC, GLOBALFOUNDRIES, and SMIC specialize in manufacturing complex chips and in designing and maintaining the latest semiconductor processes.

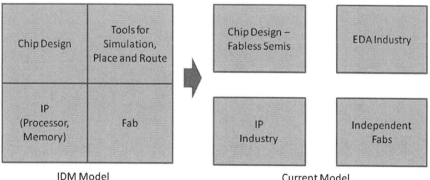

Figure 1-2 How the semiconductor industry has evolved.

This repartitioning of the chip design process shows the basic principles of divide and conquer in action. When systems become too complex, it is more efficient to divide the system into several smaller components, with a formalized interface between them. These interfaces are not free, though. Interfacing to an outside fab house is not trivial: the design files that one delivers to TSMC must meet TSMC's specifications and requirements. It also costs a significant amount of money to build chips at a commercial fab. But as the cost and complexity of fabricating chips rose, the cost of the interface became worthwhile. That is, TSMC (and others) hide the complexity of the manufacturing process from its customers and provide a relatively simple, formalized interface that allows designers to create chips without knowing the details of the manufacturing process.

This decoupling effect is the key to good interface design. The partitioning of the system, and the design of good interfaces between its components, results in transforming a large, complex, flat problem into a set of small problems that can be solved locally. It decouples the complexity of one component from the complexity of another.

The different tasks involved in digital design have also evolved over the years. In the late 1980s, synthesis was introduced, using tools to optimize digital circuits. In the 1990s, design reuse became a common methodology as chip designers realized it was more efficient to buy certain common components - such as processors and interfaces - and focus their design efforts on differentiating blocks.

In recent years, the verification of complex intellectual property blocks (IP) and SoC designs has become so challenging that specialized verification languages and verification engineers have emerged as key elements in the design process.

The process of designing complex chips is continually evolving. At each step, the industry has addressed the increasing complexity of design by separating different aspects of design, so that each aspect or task can be addressed independently. RTL and synthesis technology allows us to describe circuits independently from a specific technology or library. Design reuse and IP allow us to separate the design and verification of complex blocks from the design of the chip itself. Recent improvements in verification technology have occurred as design and verification have become separate functions in the design team.

The General Model

In general, then, the design of a complex system consists first of decomposing it into component parts. Figure 1-3 shows a cell phone as an example.

The phone contains a printed circuit board with a (small) number of key chips. Each complex chip (SoC) consists of a number of subsystems. Each subsystem consists of a number of blocks – either original designs or IP. Each block in turn consists of a number of subsystems or layers, and each subsystem consists of a number of modules.

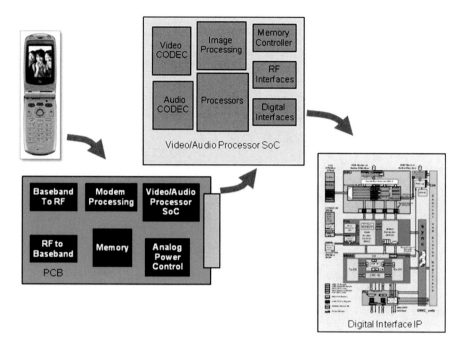

Figure 1-3 Cell phone and its component parts.

The module is the leaf level of this hierarchy. This is where the detailed design is done. It consists of HDL code describing the detailed function of the design.

Note then that there are multiple levels of design:

- Product design (the cell phone)
- PC board design
- SoC design
- IP design

At each level of design, we decompose the design problem into a set of components. For instance, at the SoC design level, we define the functional units of the design and the IP we will use for each functional unit. Then we must compose these units into the system itself. That is, we need to decide how to connect the IP together (at the SoC design level) or how to connect the modules together (at the IP design level).

At every level of design, this composition function consists of designing the interfaces between units. The design of these interfaces is one of the key elements in controlling the complexity of the design. For a good design, the design units (modules, IP, or chips) must be designed to have interfaces that isolate the complexity of the unit from the rest of the system. We will talk much more about this throughout the book.

The general paradigm then is shown in Figure 1-4.

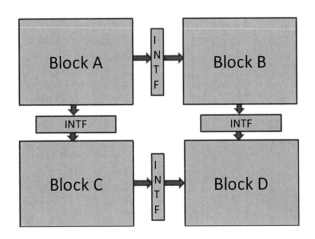

Figure 1-4 A general design paradigm for systems.

Rule of Seven

At any level of hierarchy, the number of blocks in the design is not arbitrary. In the 1950s, The psychologist George Miller published a paper called *"The Magic Number Seven Plus and Minus Two"*[1]. In this paper, he demonstrated that the human mind can hold seven objects (plus and minus two) at any one time. This is why telephone numbers (at least in the United States) are seven digits. We can remember seven digit phone numbers. We cannot remember 12 digit phone numbers.

Similarly in any design, at any level of hierarchy, we can at any one time understand a design of up to seven to nine blocks.

Compare the two block diagrams in Figure 1-5. In the diagram on the left we see only nine high-level components. It is easy to understand the major components of the system and how they are related to each other.

In the diagram on the right, much more detail is shown. This makes it significantly more difficult to understand the general functions. When we look at the diagram on the right we tend to focus in on one subsystem at a time, and try to understand what it does. Then we look at the larger diagram to understand the general, high-level functionality.

This is a common problem in design. To design effectively, we need at any one time to be looking at only a small number of design objects. By doing so, we can think effectively about the (sub)system we are designing.

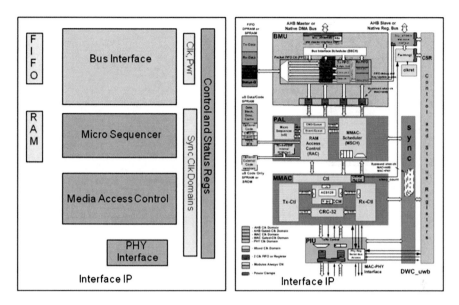

Figure 1-5 Two views of the same interface IP.

Tightly Coupled vs. Loosely Coupled Systems

Once we have partitioned the system or design into an appropriate number of design units, the key step is to design the interfaces between them. In general, systems can be considered to be one of two types - tightly coupled or loosely coupled - based on what kind of interfaces they have.

Tightly coupled systems have interfaces that essentially connect the elements of the system into one single, flat unit.

One example of a tightly coupled system is the weather. In 1961, Edward Lorenz was doing computer modeling of weather systems. He decided to take a short cut, and entered .506 for one variable, where earlier he had used .506127. The result was a totally different weather pattern.

Later, he published a paper describing this surprising effect. The title was *Does the flap of a butterfly's wings in Brazil set off a tornado in Texas?*[2]

His key discovery was that small, local effects can have large, global effects on the weather. This is typical of a tightly coupled system. Local causes can have global effects and local problems can become global problems.

In SoC design, a tightly coupled system is one where the interfaces between units create such tight interaction between the units that they essentially become a single flat design. Thus, a change to any unit in the design may require changes to all the other units in the design. Also, fixing a bug in anyone unit may require significant changes to other units.

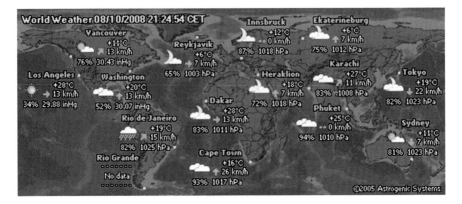

Figure 1-6 World-wide weather is a tightly coupled system. Copyright Astrogenic Systems. Used by permission.

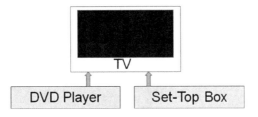

Figure 1-7 Example of a loosely coupled system.

A loosely coupled system, on the other hand, uses interfaces to isolate the different units in the design. An example of a loosely coupled system is a home video system, such as shown in Figure 1-7.

Although the DVD player, the set-top box, and the TV set are all very complex systems, the interface between them effectively isolates this complexity. The HDMI cable that connects the set-top box to the LCD display isolates the complexity of the set-top box from the display. The display does not need to understand anything about the internal behavior of the set-top box. All it has to do is understand the signals coming over the HDMI cable.

More importantly, if the set-top box breaks, we can still use the DVD player. Local bugs or defects in the set-top box do not become global problems for the entire system.

There are significant advantages to loosely coupled systems. But there are advantages to tightly coupled systems as well. Tightly coupled systems can be more efficient. This is one of the reasons they occur in nature so often. It is also one of the reasons why many design teams prefer to do place and route flat rather than hierarchical.

The characteristics of a tightly coupled system are:

• they can be more efficient than loosely coupled systems
• they allow for global optimization

- less planning is required
- they are harder to analyze
- local problems can become global problems
- they can result in "emergent" behavior: that is, big surprises

Characteristics of a loosely coupled system are:

- they are more robust
- they support the design of larger systems
- they are easier to analyze
- local problems remain local
- they require more design and planning effort
- they produce more predictable results
- they are easier to scale – they can become larger without becoming excessively complex

Flat place and route provides an excellent example of the characteristics of a tightly coupled system. A flat place and route allows optimization of the entire design, resulting in a denser, more area efficient layout. Less floorplanning is required. But a last minute ECO or bug fix can cause major problems. If the layout is so dense that a few more gates cannot be inserted where needed, then the whole place and route must be re-done from scratch. The local problem of adding a few gates has become a global problem of redoing place and route for the entire chip.

A hierarchical place and route requires much more floorplanning, and inevitably leads to less optimal density. But if one block needs an ECO or bug fix, there is a much higher likelihood that only that block will be affected; the physical design of the other blocks can remain unchanged. Local problems remain local, and the chances of a last minute bug fix causing a massive disruption are much lower. This in turn leads to a more predictable final product cost and project schedule.

One comment on emergent behavior: the academic discipline of complex systems theory has devoted a lot of effort to studying naturally occurring, tightly coupled systems. These scientists study complex systems in biology, sociology, economics, and political science. They have found that complex (tightly coupled) systems frequently exhibit surprising and unpredictable behavior. It is not possible to analyze tightly coupled complex systems by decomposing them into components and analyzing the behavior of individual components. So it becomes impossible to come up with a simple model that can be analyzed and simulated effectively. The result is that analysis is always partial, and the behavior of the system is never completely understood.

One classic example of this is an anthill. The behavior of an individual ant is extremely simple. But an entire colony of ants can exhibit quite complex behavior, including complex strategies to locate and acquire food, complex strategies for defending the colony against invaders, and in extraordinary circumstances even moving the entire colony. There is no way we could predict this kind of behavior from an analysis of the individual ant.

For more discussion of complex systems, see *Complex Adaptive Systems* [3a] and *Unifying Themes in Complex Systems* [3b].

The lesson for designers of SoC's is that very large, tightly coupled systems are much more likely to exhibit unexpected behaviors than loosely coupled systems. Such systems are difficult (if not impossible) to understand completely and to verify completely. In particular, tightly coupled hardware systems are more likely to fail in unexpected and catastrophic ways than loosely coupled systems.

Thus, one of the keys to the good design of SoC's is to make sure that they are loosely coupled systems.

The Challenge of Verification

One of the biggest challenges in digital design today is achieving functional correctness. As designs become larger and more complex, the challenge of functional verification has become extremely difficult.

In particular, a number of new, sophisticated verification techniques have been added to the engineer's toolbox: constrained random testing, assertion-based testing, score-boarding, and formal verification. Unfortunately, these new techniques merely extend the previous trend in verification. We know that with more effort (and more sophisticated tools) we can find more bugs. But we have no useful model for how to find all of them.

A quantitative analysis is compelling. There is little reliable data for bug rates in RTL code, but there is a very large amount of data on software quality. Since in both cases we are using code as a means of design, it is likely that we can extrapolate some useful information from the software quality studies.

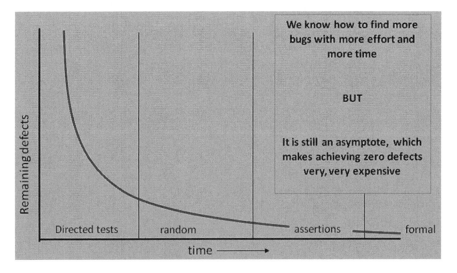

Figure 1-8 Verification is an asymptote.

Software studies show that designers inject approximately 1 defect for every 10 lines of code. The process of compiling, reviewing, analyzing, and testing the code reduces this rate. For typical new code that has been well tested, a remaining defect rate of several defects per thousand lines of code (KLOC) is common. The very best code from the most sophisticated teams can achieve between 0.5 and one defect per KLOC [5].

One exception to this quality level is NASA. NASA has been able to achieve remarkable defect rates on mission-critical software, such as for the space shuttle: perhaps as low as.004 defects per line of code, but at a cost of about $1,000 per line of code [24](compared to about $25 per LOC for commercial software).

It is likely that hardware teams commonly achieve the same kind of defect rate as the very best software teams. One reason for this is that hardware teams typically invest significantly more resources in verification than software teams do. But all the evidence indicates that any code-based design of significant size will still have a significant number of residual defects even after the most rigorous testing effort – short of adopting the cost structure and schedule of NASA projects.

As we build more and more complex chips, even very low defect rates such as 0.1 defect per KLOC can be a major problem. In fact, functional bugs are a large, if not the largest, single cause of chip respins [6a][6b].

For example consider a 50 million gate design – that is, one with 50 million logic gates. This corresponds to approximately 10 million lines of code. At 0.1 defect per KLOC, this means that the chip is likely to ship with 5,000 defects.

There are two primary strategies for improving the situation:

- reduce the number of lines of code
- lower the defect rate per thousand lines of code

In this book, we will discuss design techniques for implementing both of these strategies by raising the level of abstraction in design.

The Pursuit of Simplicity

Consistently, studies indicate that the design reviews and code reviews are the most productive means of detecting bugs[7]. Even today, the most powerful verification tool in our toolbox is the human mind. One key element in lowering the defect rate in RTL code is to make the code easier for humans to understand.

Today, most RTL code has been written with the primary goal of making it easy for the compiler (synthesis tool) to produce good results. However, this has often resulted in code that is difficult to understand. All of us have written code, only to come back to it months later and find that we have no idea what the code actually does. Unfortunately, code that is difficult to understand is also difficult to review and for humans to detect bugs.

One of the goals of this book is to describe some techniques for simplifying RTL designs (by minimizing the state space) and to simplify the way this design is represented in the code itself.

The Changing Landscape of Design

Another reason for this pursuit of simplicity is that chip design has changed in a fundamental way over the last decade. A dozen years ago, it was possible for a single senior engineer to understand every aspect of the chip -- the overall architecture, what each module does, and how software interacts with it. In a finite amount of time, a single reviewer could read all of the RTL for a chip.

Those days are gone. Today an SoC consists of many pieces of IP, often purchased from multiple third-party IP providers. It also requires large amounts of software, from low-level drivers to high-level applications. The RTL for a complicated IP may be as large as an entire chip was just a few years ago.

As a result, no one engineer completely understands every aspect of an SoC design. No single human being could read all of the RTL for such a design. The design, verification, and debug a complex chip now requires an entire network of engineers.

As a side effect, during verification and debug, engineers are constantly dealing with code that they did not write, and which they may never have looked at before. They may not be experts on the bus protocols used on the chip or the interface protocols used in the I/O for the chip.

Another way to look at this paradigm shift is this: in previous generations of design, chips were designed by teams of engineers. Now they are designed by a network of engineers. Managing this network is much more complex than managing a team.

With a team, any member who has a question has direct access to the other team members and can find the answer to the question quickly and directly. With a network, an engineer may have no direct access to the engineer who can answer the question. Access may require multi-node hops: finding where a IP came from, locating the field contact for the IP company, getting the field contact to relay the question to the design team, and so on. With a network, more information may be available, but accessing it quickly can be more difficult.

In this new model for chip design, there is a premium on simplicity: a simple, regular architecture, robust IP's that are easy to integrate, and code that is easy to read and understand.

Structure of This Book

Chapter 2 gives a brief overview of techniques for simplifying designs.
Chapter 3 gives a detailed example of how to re-factor an RTL design to make it significantly less complex. It uses a control dominated design as the example.
Chapter 4 continues the example of Chapter 3, focusing on the design of a hierarchical state machine.
Chapter 5 discusses in more detail the concept of state space, and how to minimize it.

Chapter 6 describes how the techniques described in previous chapters can simplify and improve verification.

Chapter 7 gives another detailed example of simplifying code, this time with a data path intensive design.

Chapter 8 describes how the design of module interfaces affects the complexity of the design.

Chapter 9 continues the discussion in Chapter Chapter 7, and extends it to the IP and system level; it describes how to measure and minimize complexity of complete designs.

Chapter 10 begins a discussion of raising the level of abstraction by extending current design languages and tools.

Chapter 11 describes a series of proposed extensions to SystemVerilog that could start moving RTL up in abstraction.

Chapter 12 discusses the future of design – the potential for greatly improving designer productivity and the challenges and obstacles to realizing this potential.

Appendix A summarizes some of the design guidelines developed in the course of the book.

Appendix B provides some code examples of designs using the recommended coding styles for SystemVerilog as well as some examples of designs using the proposed extensions described in Chapter 10.

Appendix C provides some preliminary specifications for the proposed extensions to SystemVerilog.

Appendix D discusses some existing SystemVerilog features that can be useful in raising the abstraction of code.

Chapter 2
Simplifying RTL Design

*Confusion and clutter are the failure of design, not the attributes
of information.*

—Edward R. Tufte

This chapter gives an overview of the challenges in RTL designs, and some of the
basic techniques we can use to simplify them.

Challenges

The basic challenge in RTL design is that there are a lot of things going on at the
same time. The design of hardware involves dealing with concurrency. And currency
is inherently a difficult problem.

In addition, in RTL we describe both the function of the design and a great deal
of the implementation details. For instance, we define the basic clocking structure
and whether reset is synchronous or asynchronous. By the way we write the RTL
we determine whether latches or flip-flops will be used.

Historically, we have used code structure and coding style to develop code that
is synthesis friendly, easy to achieve timing closure, and meets our power and gate
count constraints. Clarity of the code has often been a secondary concern.

As designs become more complex, the challenge of describing both function and
implementation at the same time becomes even more difficult. For instance, interface protocols such as USB 3.0 involve a number of complex algorithms. Although
we think about these algorithms as operating on packets, these are serial interfaces;
we must implement the algorithms serially, operating on one bit or one word at a
time. Developing the correct algorithm and at the same time defining its serial

---

M. Keating, *The Simple Art of SoC Design: Closing the Gap between RTL and ESL*,
DOI 10.1007/978-1-4419-8586-6_2, © Synopsys, Inc. 2011

implementation is a complex task. As in any complex task, at some point it becomes easier to divide it into two separate tasks, and solve them separately.

One of the byproducts of designing both the function and the implementation details simultaneously is that the code size tends to become quite large. Source code file sizes can often run into the tens of pages. The code tends to be structured to be friendly to the compilers not necessarily to the humans who read and debug the code. All this results in code that is difficult to analyze, review, and debug.

Syntactic Fluff

Another byproduct of trying to write synthesis friendly code is that we end up with a lot of syntactic fluff. For example, describing a simple flop might consist of the following code:

```
always @(posedge clk or negedge reset) begin
  if (!reset) foo <= 0;
  else foo <= foo + 1;
end
```

In this case, the only part of the code that is algorithmically significant is the line:

```
foo <= foo + 1;
```

The rest of the code is syntactic fluff. That is, it is required in order to convince the synthesis tool that a flip-flop should be used and tell it the nature of the clock and the reset signal as well as the reset value of *foo* (which is zero for most flops).

Another example of writing synthesis friendly code is the practice of separating the code into combinational and sequential sections. In the early days of synthesis, we could get better results by putting all the combinational code at the beginning of the file and all the sequential code at the end of the file. So code might look something like the following:

```
assign a = b;

always @(c or d) begin
  e = c && d;
  f = c || d;
```

(continued)

(continued)

```
      end
      always @(posedge clk or negedge resetn) begin
        if (!resetn) foo <= 0;
        else foo <= a;
      end

      always @(posedge clk or negedge resetn) begin
        if (!resetn) bar <= 0;
        else bar <= e + f;
      end
```

This structure, of course, makes no logical sense. Logically, the combinational code that defines the value of *a* should be right next to the sequential code where *a* is used.

With today's synthesis tools, this kind of partitioning provides no value at all. The synthesis tools can optimize all the code across a very large module regardless of how the code is organized or structured.

One of the themes of this book is that we need to migrate our coding style from being synthesis friendly to being human friendly. The synthesis tools have become much more sophisticated over the last 10 years, but at the same time the designs have become much more complex. As a result, we have an opportunity to rethink how we code digital designs make them easier to understand and analyze. The power of modern synthesis tools gives us a lot of leeway to modify how we write code in order to make the design process faster and more robust.

Concurrency and State Space

There are several problems in RTL design that are simply the result of how hardware description languages and synthesis tools evolved. This category includes syntactic fluff and the fact that we describe function and implementation in the same file.

But there are two major challenges in RTL design that are fundamental to the problem of digital design: concurrency and state space. These two issues are closely related.

When we design a digital system, we are really specifying how that system evolves over time. That is, we are specifying the state space of the system and how it changes over time. The problem is that the state space may be very complex, consisting of multiple subsystems that are evolving simultaneously.

Consider, for example, a cell phone. The main digital chip in a cell phone may be simultaneously controlling the user interface, the audio and video services, network access, and the radio subsystem.

We can demonstrate the challenge of such complex systems from a very simple example. Consider the state machine in Figure 2-1.

Note: In this book, we use a mix of styles in state machine diagrams. For very simple diagrams, we use traditional bubble diagrams. For state machine drawings where we show some code, we use State Chart notation. This format (using rectangles instead of circles for states) gives room for including more information about the state. For an explanation of this format, see [11].

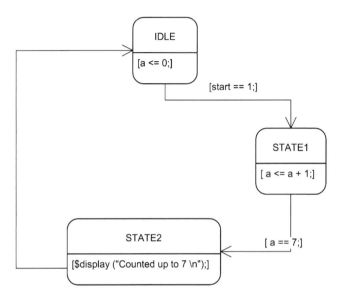

Figure 2-1 A simple state machine.

Analyzing the state machine is quite simple. We just have a counter that counts up to seven once the start signal is asserted.

If we have two state machines that are decoupled, as in Figure 2-2, the analysis is again simple:

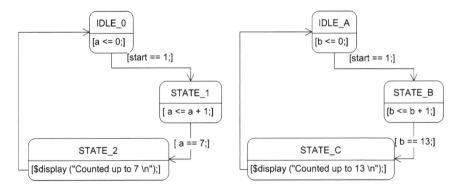

Figure 2-2 Two decoupled state machines.

Now we have two state machines that count up to some terminal value, starting when the start signal is asserted. Note that because the two terminal counts are relatively prime, there is no way to predict the value of b given the value of a. After a hundred clock cycles or so, the relationships between the values of a and b will appear to be completely random. Thus, while it is easy to analyze each state machine independently, analyzing and predicting the values of both states at any particular time starts to get a bit tricky.

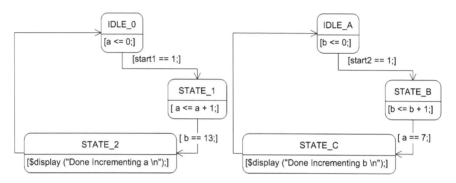

Figure 2-3 Two coupled state machines.

In Figure 2-3 things are getting dicey. In the above design, the two counters have separate start signals. Also, we halt incrementing a based on the value of b, and vice versa. The two state machines are now tightly coupled, and the combined behavior depends heavily on when the two start signals are asserted. The behavior of this circuit is a lot more complex than the behavior of the previous two circuits.

As we can see, the concurrent behavior of two tightly coupled state machines can become very complex to analyze, even when each state machine is simple.

Techniques

The previous sections described three problems in RTL design:

- Syntactic fluff
- The order/structure of RTL code
- The problems of state space size and complexity, and the problem of concurrency

We now give a brief overview of some of the techniques we can use to address these problems. These techniques will be explored in more detail in the rest of the book.

As mentioned earlier, the key technique for managing complexity is to divide and conquer. In terms of RTL design, and in fact in any code based design, the key mechanism is encapsulation. We want to partition the design – and the code – so that each piece can be designed and analyzed separately from the other pieces. To the degree possible, we would like to encapsulate functionality, hide local information so that external pieces of the design don't see it, and present a simple interface to the rest of the system.

Even with today's languages and tools, we can use encapsulation techniques to raise the level of abstraction above the traditional RTL level. In doing so, we can make the function of the design more obvious and make the implementation less obtrusive.

In this section, we will examine four areas for encapsulation and raising the abstraction level of design:

- Combinational code
- Sequential code
- Interfaces
- Data Types

Encapsulating Combinational Code

Consider the following piece of SystemVerilog code:

```systemverilog
input bit a;
input bit b;
input bit control;

bit temp;
bit [7:0] foo;

always_comb begin
  if (control == 1) temp = a;
  else temp = b;
end

always_comb foo = temp * 3;
```

In this case, the signal *temp* has global scope. That means that when we are analyzing this design, we need to worry about the value of *temp* at all times. But in fact, the signal is used only as a temporary or intermediate value in calculating *foo*.

Compare the previous counter to the following code:

```
function automatic bit [7:0] foo (input bit a, b,
control);
    bit temp;
    if (control == 1) temp = a;
    else temp = b;
    foo = temp * 3;
endfunction
```

This code is slightly shorter than the previous code. But it also has several additional advantages:

1. It makes it completely explicit that the value of *foo* depends only on the inputs *a*, *b* and *control*. This relationship is not at all obvious from the statement *always_comb foo = temp * 3*. In fact, if the two *always_comb* blocks in the previous example are separated by significant amounts of code, it may not be easy at all to see the relationship between *foo* and the inputs *a, b*, and *control.*
2. The signal *temp* is local within the function. It is completely obvious that it is not used by any other piece of code.
3. All of the code required to calculate *foo* is grouped together within the function. There is no possibility of scattering this code throughout the file. This means that the analysis of how *foo* is calculated becomes a local rather than a global activity.
4. The function *foo* must now be called explicitly whenever it is needed. This makes coding slightly more burdensome, but it makes analysis significantly easier. Typically, the function will be called in one or perhaps a few states. That means whenever the module is in the other states, we can completely ignore *foo*.

Thus, functions provide an effective encapsulation mechanism for combinational code.

Structuring Sequential Code

Unfortunately, modern hardware description languages do not provide an equivalent encapsulation mechanism for sequential code. There is no structure that allows us to group pieces of sequential code together, define explicitly the inputs, or to hide local or temporary signals. The *task* construct allows some degree of encapsulation, since (unlike *function*) it allows some timing and sequential constructs. And we will use it in a later chapter. But we are not allowed to have an *always @ (posedge clk)* block in a task. As a result, we really do not have an equivalent to the *function* for sequential code.

Instead, we are left to group sequential code arbitrarily within *always @ (posedge clk)* blocks. These sequential blocks can be scattered throughout a file. To analyze the module then, it is necessary to read and memorize virtually the entire file

Consider the following code:

```
always @ (posedge clk or negedge resetn) begin
  if (!resetn) begin
    bar <= 0;
    bar_p1 <= 0;
  end else begin
    bar_p1 <= bar;
    bar <= a + b;
  end
end

always @ (posedge clk or negedge resetn) begin
  if (!resetn) begin
    foo <= 0;
  end else begin
    foo <= bar_p1 + bar;
  end
end
```

Here it is not obvious that *foo* depends on the inputs *a* and *b*. If the two sequential blocks are separated by significant amount of code, it may be nontrivial to sort out exactly what the relationship is between *foo* and *bar*.

One possible solution is to start grouping more and more sequential code into a single sequential process. The trouble with this solution is that this process becomes large and unwieldy.

The best mechanism for structuring sequential code is the state machine. In a state machine, we can create a single large sequential process that uses the case statement to structure the sequential code into separate states.

To address the problems of concurrency described earlier, we recommend using a single state machine per module. Effective decoupling of modules (described in Chapter 8) then helps manage concurrency between state machines.

The key challenge in grouping large amounts of sequential code into a single state machine is that this state machine can rapidly become large and unwieldy itself. In fact, we can easily violate the rule of seven: many interesting state machines have more than seven to nine states. The solution to this problem is to code the process as a hierarchical state machine. We discuss hierarchical state machines Chapter 4, and give an example in Appendix B.

Using High Level Data Types

Functions and state machines are the two most important mechanisms for encapsulation in RTL design. But there are some additional techniques available in SystemVerilog that can be very helpful in raising the abstraction level of RTL design.

Enumerated types are helpful in defining exactly what values are legal for a given signal or collection of signals. For instance:

```
bit read;
bit write;
```

This code implies that there are four possible values for the combination of the read and write signals. Most importantly, it implies that it is possible to assert both read and write at the same time; at least nothing in the declaration implies that this is impossible.

Instead, we can define an enumerated type signal *rw* which makes it explicit that only one of the read or write operations can be active at one time:

```
enum (NOP, READ, WRITE) rw;
```

Structs in SystemVerilog are also very useful in providing an encapsulation mechanism for related signals. For instance:

```
bit [ADDR_WIDTH] foo_address;
bit [ADDR_WIDTH] bar_address;

enum (NOP, READ, WRITE) foo_rw, bar_rw;

bit [DATA_WIDTH] foo_data;
bit [DATA_WIDTH] bar_data;
```

As written, the code relies on the signal name to imply the relationship between the different signals.

```
typedef struct {
  bit [ADDR_WIDTH] address;
  bit [DATA_WIDTH] data;
rw_type rw;} my_data_type;


my_data_type foo, bar;
```

Using a *struct* data type, we can make it explicit that both foo and bar are exactly the same data type, with exactly the same type of address, data and control signals. The relationship between the address, data, and control signals is much more explicit as well.

The SystemVerilog *interface* construct provides an encapsulation mechanism at the interface level. A module definition with 30 or 40 inputs and outputs clearly violates the rule of seven. Using the interface construct, we can reduce this to seven to nine interface declarations.

The following is an example of how a simple memory interface can be defined using interfaces:

```
interface mem_intf ; // interface for i_mem and d_mem
  bit [ADDR_WIDTH-1:0] addr;
  bit [WORD_SIZE-1:0] write_data;
  bit [WORD_SIZE-1:0] read_data;
  bit read;
  bit write;

  modport master (output addr, write_data, read, write,
                  input read_data);
  modport slave (input addr, write_data, read, write,
                 output read_data, exc );

endinterface: mem_intf
```

Then in the top level module, we instantiate an interface and connect it to the memory. Note how simple the code for the instantiating the memory has become, since only the interface, and not five different ports, needs to be connected.

```
module top ;
   ...
mem_intf d_mem_intf();
   ...
mem d_mem (.ifc(d_mem_intf), .clk(clk));
   ...
endmodule
```

Then our behavioral model for the memory might look something like this. Note how simple the port declaration has become, since we declare the interface instead of five different ports.

```
module mem (input bit clk, mem_intf ifc);

   bit [`WORD_SIZE-1:0] mem_array [`MEM_DEPTH-1:0] ;

   always @(posedge clk) begin
      if (ifc.read) ifc.read_data <= mem_array[ifc.addr];
      if (ifc.write)mem_array[ifc.addr] <= ifc.write_data;
   end
endmodule
```

For an extensive discussion of how to use the *interface* construct, see [8]. For a brief discussion of how extensions to the synthesizable subset of SystemVerilog could make the interface construct even more useful, see the first section of Appendix D.

Finally, even the *for* loop now has a small opportunity for encapsulation:

```
for(int index = 0; index < max_val; index++)
```

By declaring the loop index inside the for loop, we hide it from the rest of the code.

Thinking High-level

Most important of all, raising the level of abstraction of RTL code requires us to think high-level in every aspect of coding. For example, consider the following piece of code:

```
if (foo == 1'b1)
```

This is an example of thinking at the bit level. We are asking if the value of *foo* is equal to one, which we associate with a Boolean value true.

The following piece of code is functionally the same as before, but simpler and at a higher level of abstraction:

```
if (foo)
```

In this statement, we simply ask if foo is true. In fact, we know that this is equivalent to asking if foo is not equal to zero.

There are several (admittedly small) problems with the first approach.

1. It is more verbose than necessary, which can become a significant issue when reading large amounts of code.
2. It inserts an implementation issue (the fact that we are using a value of one represent a Boolean value true), when we are really interested in the functional or algorithmic aspects of the design.

Both ways of writing an *if* statement are perfectly legal, and both will produce exactly the same synthesis results, that is, the same gate level netlist. But the second version is more compact and more functional rather than structural.

All the techniques described in this chapter strive to achieve a single goal. There are many different ways of writing the same logic in RTL code. In the past, we had to choose the coding style that lead the synthesis tools to produce the optimum result. But today, with the explosion of complexity in design, we need to use a coding methodology that makes the code easy to understand, to review, to analyze and to debug.

Chapter 3
Reducing Complexity in Control-Dominated Designs

> *The ability to simplify means to eliminate the unnecessary so that the necessary may speak.*
>
> — Hans Hofmann

This chapter provides an example of how to reduce the complexity of an RTL design. It describes a project to rewrite a module and raise the level of abstraction without changing its functionality.

The project started with the Bus Control Unit (BCU) from a wireless USB design. This module initiates and controls the DMA from the MAC layer to the AMBA® AHB bus, as shown in Figure 3-1.

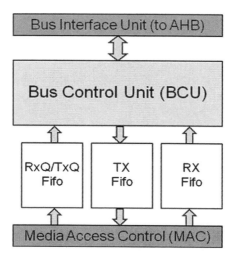

Figure 3-1 Simplified block diagram of the wireless USB design.

M. Keating, *The Simple Art of SoC Design: Closing the Gap between RTL and ESL,*
DOI 10.1007/978-1-4419-8586-6_3, © Synopsys, Inc. 2011

We used the following metrics to assess our success in simplifying the RTL code:

- number of lines of code
- internal state space of the design
- number of objects in the design
- amount of abstraction, partitioning, and information hiding in the design
- the overall readability of the code

One constraint of the project was that we could not change the interface of the module. To test the design, and prove that we had not changed its function, we needed to be able to interface to the other blocks in the existing wireless USB design. For this reason we are not addressing the issues of input and output state space size.

The techniques that we used to simplify the code include:

- partitioning the code into multiple files
- eliminating some syntactic fluff from the code
- restructuring the sequential code, moving it all into a single state machine
- using functions for the combinational code
- converting the state machine to a single, sequential process
- converting the state machine to a hierarchical state machine
- using the SystemVerilog *struct* construct to encapsulate important structures

Original Code

The initial design was chosen because it was one of the largest modules in the design and reasonably complex. It was also well-designed and well-coded; that is, it was designed and coded in compliance with the best common design practices. Any improvements would have to move beyond these practices and possibly require changes to the Verilog language.

The original Verilog-2001 code consisted of twenty-seven pages for a total of 1600 lines of code (total lines of text including comments and blank lines), including

- 2 pages of input and output declarations
- 4 pages of internal declarations (wires, regs, parameters)
- 7 sequential processes (always @(posedge clk))
- 10 combinational processes (always @*)
- 30 assign statements

The sequential processes included :

- One process that was 5 pages long (a state machine)
- second process that was 3.5 pages long
- third process that was 1 page
- The rest of the sequential processes were less than 1 page each

The combinational processes included :

- One process that was 2 pages long
- One process that was 1 page long
- The rest of the processes were less than 1 page

Thus, of the 27 pages,

- 6 pages (22%) were declarations
- 12.5 pages (46%) were large processes

These statistics are useful because declarations, although certainly necessary, are largely a distraction when trying to read and debug code. Large processes (more than 1 page) are more difficult to read and understand than small processes. A state machine that takes 5 pages is very challenging to read and understand.

The rewrite project consisted of three steps:

- Analyze and understand the code
- Rewrite the code
- Test the new code in the wireless USB test environment

During the analysis phase, we did not consult with the original designer. We wanted to see how easy (or difficult) it was to understand the function of the design just from the code and a very high-level specification document. (This specification document, of course, was ambiguous and contained a lot of obsolete information. Like most specifications, it was not updated as the design changed).

During testing, we were not able to use formal verification to show that the new version and old version were equivalent. As we restructured the code, we converted the state machine into a single sequential process from the original two-process format. As a result, the two designs had slightly different timing – that is, certain internal actions moved from one clock cycle to another. (We could have maintained exactly the same behavior, but the resulting code would have been more complex, undermining the primary goal of the exercise). This rendered the two designs non-equivalent (from the perspective of formal verification) but functionally equivalent, in terms of passing all the regression tests.

State Space in the Original Design

In the original Verilog code, the state machine consisted of 12 states.

The next two largest sequential processes assigned values to 44 registers and 9 registers respectively. On first reading the code, it was not obvious which combinations of these 53 registers can occur and which are impossible. Thus, the state space of these two processes was 2^{53}.

Without more information, we had to assume that any of these 2^{53} states could occur during any of the 12 states of the state machine.

Thus, the total state space, just of the state machine and these two processes, was $12 * 2^{53}$. In human terms, this is essentially infinite.

The original designer, of course, intended a much smaller state space. In fact, the original designer would see most of the variables in the sequential processes as data or temporary variables rather than control states. Because of his understanding of the design intent, the effective state space for the original designer is much smaller than the state space as actually represented in the code.

During the analysis phase of the re-write project, one key goal was to restructure the code to make it easier to understand the state space. We could then rewrite the code to minimize the state space and make it explicit and obvious.

Partitioning

The first step in restructuring the design was to partition it into several files. This allowed us to aggregate related code into separate code segments that could be analyzed and rewritten independently.

We partitioned the design into the following files:

- input and output declarations
- internal declarations of wires, registers and parameters
- all combinational code (with a few exceptions, explained below)
- all sequential code (with a few exceptions, explained below)
- the top level code, consisting of:

  ○ the module declaration
  ○ '*include* statements to include the other files
  ○ combinational code that directly drives primary outputs
  ○ sequential assignments that simply pipeline primary inputs

With this partitioning, virtually all the real action of the module was contained in two files: the combinational code and the sequential code. During the subsequent work we could largely ignore the other files, referring to them only as required.

Comments

The next step was to restructure the comments in the code.

Note that the comments in Example 3-1, although they contain some interesting information, completely interrupt the flow of the code. Scattering comments throughout a process actually makes reading the code harder. Instead, we moved the useful comments ahead of the process, so all the comments could be read at once, and then all the code could be read without interruption.

```
// Flip Flop registers for Byte Stripping
always @ (posedge hclk or negedge hreset) begin
  if (!hreset) begin
    rxf_saved_data <= 0;
    rxf_saved_data_cnt <= 0;
  end else begin
    if (fsm_rxqreq_vld_clr ||
        (fsm_q_sel == DWQ)))) begin
      rxf_saved_data_cnt <= 0;
    end else if (fsm_rxf_saved_data_load ||
                 (bium_wdata_pop &&
                 (rxf_data_shift != 0))) begin
// bium_fifo_busy not needed - no pop on BUSY
// load RXF data register if:
// - FSM instructs to do so
// - or BIUM pops RXF and byte shifting is
// required
      rxf_saved_data <=
        bcu_rdata[`DWC_UWB_BUS_DWIDTH-1:8];
// determine number of valid bytes saved in RXF
// data register based on
// shift value
      case (rxf_data_shift[1:0])
        2'b00: rxf_saved_data_cnt <= 'h0;
        2'b01: rxf_saved_data_cnt <= 'h3;
        2'b10: rxf_saved_data_cnt <= 'h2;
        default: // 2'b11
              rxf_saved_data_cnt <= 'h1;
      endcase
    end else if (bcu_rxf_pop || bcu_dwf_pop ||
                 bcu_rxf_pktcnt_dec ||
                 (fsm_dmareq_vld_clr &&
                 (dmareq_f == DWF))) begin
// clear num valid bytes if we don't load
// RXF data register this clock and:
// - BCU pops RXF (last entry)
// - or BCU decrements RXF packet counter
// (upon DMA completion or RX pkt flush)
      rxf_saved_data_cnt <= 0;
    end
  end
end
```

Example 3-1

We ended up eliminating many of the comments, because they simply restated what the code says. We kept only the comments that explained particularly tricky pieces of code.

Engineers are taught in school to use comments extensively. We find comments to be problematic: they are not tested in simulation, and they are typically not updated as the code changes. The result is that they are often incorrect and misleading. We prefer to use comments sparingly, mostly to explain the meaning of signals and to give a general ideal of the algorithm being implemented.

Syntactic fluff

Preprocessing Sequential Code

Once the design was partitioned into manageable sized files, and extraneous comments removed, we still found the code difficult to read and understand. One of the most obvious causes of this was a large amount of syntactic fluff in the code. Consider the following (very small) sequential process.

```
 1  always @(posedge hclk or negedge hreset_gen_n) begin
 2    if (!hreset_gen_n) begin
 3      csr_debug_bcu1_lo <= 4'b0;
 4      bium_abort <= 1'b0;
 5    end
 6    else begin
 7      csr_debug_bcu1_lo <= {bcu_dwf_pop, bcu_rxf_pop};
 8      if (bium_dma_done)
 9        bium_abort <= 1'b0;
10      else if (bium_abort_set)
11        bium_abort <= 1'b1;
12    end
13  end
```

This process consists of 13 lines of code. But all the meaningful action is described in lines 7 through 11. So five lines of code are of high-value, and the other seven are overhead. All they do is get in the way of understanding the code.

To make the code simpler and more readable, we developed a simple Perl script to act as a preprocessor for the RTL code. In this module, as in the vast majority of modules in the design:

- there is only a single clock and a single reset
- non-blocking assignments are used for sequential code, and only for sequential code

- blocking assignments are used for combinational code, and only for combinational code
- only flip-flops are used, never latches

With the Perl script, the sequential process above can be rewritten in a streamlined fashion as:

```
csr_debug_bcu1_lo <= {bcu_dwf_pop, bcu_rxf_pop};
if (bium_dma_done bium_abort <= 1'b0;
else if (bium_abort_set) bium_abort <= 1'b1;
```

The Perl script uses the "<=" as an indication that this is a sequential process, and re-writes the code as:

```
1 always @(posedge hclk or negedge hreset_gen_n) begin
2   if (!hreset_gen_n) begin
3     csr_debug_bcu1_lo <= 4'b0;
4   end
5   else begin
6     csr_debug_bcu1_lo <= {bcu_dwf_pop, bcu_rxf_pop};
7   end

1 always @(posedge hclk or negedge hreset_gen_n) begin
2   if (!hreset_gen_n) begin
3     bium_abort <= 1'b0;
4   end
5   else begin
6       if (bium_dma_done) bium_abort <= 1'b0;
7       else if (bium_abort_set)bium_abort <= 1'b1;
8   end
9 end
```

This code is exactly equivalent to the original code.

Note that to re-generate legal Verilog code from the streamlined code, three additional pieces of information are needed:

1. the reset values of the registers
2. the name (and edge) of the clock
3. the name (and edge) of reset

In the streamlined code, all three of these are provided in the signal declarations by the following code:

```
bit [3:0] csr_debug_bcu1_lo = 0;
bit [3:0] bium_abort = 0;

$clock posedge hclk
$reset negedge hreset_gen_n
```

That is, we use the initialization construct of Verilog to define the reset value. And we declare the clock and reset explicitly.

With this Perl script, we were able to dramatically reduce the size of the sequential code. More importantly, by eliminating the distraction of syntactic fluff, we made the code much easier to read.

Note that this approach is possible because the module in question uses only a single clock. But for the vast majority of modules in digital design, this is the case. Good design practices dictate that, whenever signals cross clock boundaries, a small separate module is used and that only this module contains multiple clocks. The Perl script approach described here does not apply to such a multi-clock module, but does apply to all the other modules in the design.

Preprocessing Combinational Code

We used a similar approach on combinational code, which is a much simpler case. Here the Perl script recognizes that a blocking assignment is used (in this kind of module) only for combinational code.

So we can reduce the following combinational process:

```
1 always @*
2 begin
3   next_txf_saved_data_dec = 0;
4   for (q=0; q<DWIDTH_LANES; q=q+1)
5     if (q < next_txf_saved_data_cnt)
6       next_txf_saved_data_dec[q] = 1'b1;
7   next_txf_saved_data_en = next_txf_saved_data_dec;
9 end
```

To this:

```
1  for (q=0; q<DWIDTH_LANES; q=q+1)
2    if (q < next_txf_saved_data_cnt)
3      next_txf_saved_data_dec[q] = 1'b1;
4    else next_txf_saved_data_dec[q] = 0;
5  next_txf_saved_data_en = next_txf_saved_data_dec;
```

The Perl script generates the appropriate legal Verilog syntax, adding back the *always@*, begin*, and *end*.

Once we had partitioned the code into multiple files, eliminated the syntactic fluff, and removed unnecessary comments, we found the code dramatically easier to read, analyze, and re-factor. We first focus on improving the sequential code.

Refactoring Sequential Code

Recoding the State Machine

There are two classic ways of coding a finite state machine (FSM):

Two-Process FSM

```
always @* begin
  case (state)
    IDLE:    if (foo) next_state = STATE1;
    STATE1:  if (bar) next_state = IDLE;
  endcase
end


always @(posedge clk) begin
  state <= next_state;
end
```

One-Process FSM

```verilog
always @(posedge clk) begin
  case (state)
     IDLE:    if (foo) next_state <= STATE1;
     STATE1:  if (bar) next_state <= IDLE;
end
```

In the original Verilog for the BCU, the state machine used the two-process approach. Although this is a common way of coding state machines, and approved in the *Reuse Methodology Manual* (RMM) [9], it can, in fact, lead to poorly structured, hard-to-analyze code. For example, we found states in the combinational state machine processes that looked like:

```verilog
STATE_XYX : begin
  if (foo) doit = 1;
```

And many sequential processes elsewhere in the code that said something like:

```verilog
always @(posedge clk) begin
  if (doit) begin
    bar <= 1'b1;
  end
end
```

In this example, using *doit* as a flag between the combinational and sequential processes makes the code more complex and harder to analyze. It adds another (sequential) process to the state machine; the state machine now consists of a combinational process, a sequential process, and a process for *doit*. There is no structure to keep these processes together in the code – they can be scattered anywhere in the module. This violates the principle of locality: related code should be located together.

If we want to find out what happens in the state *STATE_XYZ*, we have to search though the rest of the code to find all the occurrences of *doit*, and analyze the code where it appears. If there are many flags (as was the case in the BCU design) then we have to scan all the code for all the occurrences of all the flags. And these occurrences were scattered over twenty-eight pages of code.

What we would like is for all the code related to the state *STATE_XYZ* to be in one place. That would make review and analysis much easier.

In response to these observations, we converted the state machine to the one-process type of state machine, converting it to a single sequential process. This required a significant amount of work, but the result was code that was much cleaner and easier to understand. In particular, all the code (all the actions described by the code) for a particular state was in one place.

Relocating Other Sequential Code

Our next step was to examine the rest of the sequential code in the module. Our goal was to move this code from separate sequential processes (*always@(posedge clk)* blocks in the original code) into the central state machine. We found that there were basically two types of code. One type looks as follows:

```
if ((prev_state_bcu == RDQ1) && (state_bcu == RDQ2)) begin
   case (fsm_q_sel)
      TXQ: begin
         txqreq_rr[fsm_txqnum_sel] <= bcu_rdata[F_RAO_RR];
         txqreq_tag[fsm_txqnum_sel]<= bcu_rdata[F_RAO_TAG];
         ...
```

In this case, it was straightforward to move this code into the state machine, as part of the code for state *RDQ2*.

The second type of sequential code looked as follows:

```
if         (dma_clr  && (q_sel == TXQ)) req_tx <= 1'b0;
else if  (txq_clr)                             req_q  <= 1'b0;
else if  (dma_set  && (q_sel == TXQ)) req_tx <= 1'b1;
```

In this case, it's not clear whether this code can be moved into the state machine. But a careful analysis of the design indicated that the critical control signals (*dma_clr*, *txq_clr*, and *dma_set*) are just flags from the state machine.

It turned out that *dma_clr* can only be true in state *TX_STATE1*; so we just moved the statement

```
if (dma_clr && (q_sel == TXQ))req_tx <= 1'b0;
```

to the *TX_STATE1* clause in the (new) state machine case statement.

Similarly, we moved the other two if statements to the appropriate states in the new state machine.

In one or two isolated cases, a small piece of code could be active in more than one state. In these cases, a single piece of code had to be moved (copied) into two states. But in all cases, moving sequential code into the state machine made the design much easier to understand. For example, to analyze the original code, one would have to search the entire 27 page file to find all the places where *dma_clr* is referenced and then try to determine how this interacts with the rest of the state of the module.

To create the new code, we had to do exactly this analysis. But once we completed this analysis, we were able to move this code into a single state in the state machine. As a result, this code became trivial to analyze.

Using these techniques, we moved virtually all the sequential code into the state machine.

Moving all the sequential code into a one state machine, coded as a one sequential process, was the single most effective step we took in simplifying the code. As a result of moving the sequential code to the state machine, all the sequential behavior of the design could be analyzed and understood from examining a single, small file. The state machine itself became simpler, and the overall code for the module became dramatically simpler.

Rewriting Combinational Code

After restructuring the sequential code, we turned to the file that contained all the combinational code. This was a considerable amount of code and quite complex to analyze. Our major concern was how to restructure this code so it would be easy to understand. How could we hide the information that we could hide and make obvious the information that needed to be visible? In other words, what was the preferred encapsulation method for combinational code?

We decided to use functions. We ended up rewriting virtually all of the combinational code in the form of functions. For example,

```verilog
function automatic [`DWIDTH-1:0] dma_cnt (
  input [`AWIDTH-1:0] addr,
  input [`LEN-1:0]    len);

  bit [`FIFOD-1:0] fifo_depth = `FIFO_DEPTH;
  reg [1:0] stbytes;

  case (addr[1:0])
    2'b00: stbytes = 2'b00;
    2'b01: stbytes = 2'b11;
    2'b10: stbytes = 2'b10;
    2'b11: stbytes = 2'b01;
  endcase
```

(continued)

(continued)

```
  if (len <= {fifo_depth,stbytes})
    dma_cnt = len;
  else
    dma_cnt ={fifo_depth,stbytes};
endfunction
```

In this case, a great deal of the complexity in calculating *dma_cnt* is hidden within the function definition. It is explicitly stated that the function depends only on two external variables (*addr* and *len*). This kind of encapsulation can be very effective in partitioning very complex combinational code into bite sized chunks that can be analyzed separately.

To show how *dma_cnt* is used, here is a fragment of the rest of the code, starting with the main state machine:

```
always @(posedge hclk or negedge hreset_gen_n) begin
  case (dma_state)
    IDLE: begin
      if (dma_rdy) setup_dma();
      ...
    end
    ...
  endcase
end

function automatic setup_dma ()
  ...
  cnt = dma_cnt(addr, len);
  ...
endfunction
```

That is, the *IDLE* state calls *setup_dma*, which calls *dma_cnt*.

The only time we have to setup the DMA is in this *IDLE* state – in all other states, the functions *setup_dma* and *dma_cnt* are not used and can be ignored.

One concern we had about the file containing all the functions was that it was fairly large, about eight pages. The question was how to organize this code.

In the end, we made following observation. The key behavior in the module is really captured in the state machine. It describes how the behavior of the module evolves over time. On the other hand, the combinational code is really a set of definitions. For instance, the function above defines *dma_cnt* in terms of *addr* and *len*. And there is a well-established paradigm for ordering definitions, namely dictionary (alphabetical) order.

So the file of functions is simply ordered alphabetically. The function *dma_cnt* appears just after the function *ccub_txfnum* and before the function *dma_txf*.

Analyzing the New Code

Now analysis of the overall design is quite simple. One starts with the state machine. From analyzing it, we understand the basic algorithm, the basic sequence, that the module executes. In fact, the top-level algorithm is contained in the idle state of the state machine (shown in slightly simplified form):

```
IDLE: begin
  if (stop_condition)
    state_bcu <= CCUB_STP;
  else if (ok_to_read_q()) begin
    state_bcu <= RDQ1;
  else if (|dma_rdy.txf & !ccub_txf_stopped)
    state_bcu <= DMA_TX;
  else if ((dma_rdy.rxf & !ccub_rxf_stopped)|dma_rdy.dwf)
    state_bcu <= DMA_RX;
end
```

That is, if the module is told (by an external resource) to abort the DMA, we do the abort (initiated by state *CCUB_STP*). Otherwise, if there is a DMA request in one of the queues, we read the queue, initiated by state RDQ1. Otherwise, if there is a DMA request pending (as a result of reading the queue) then perform the DMA. There is a clear priority to the DMA: TX(transmit) has priority over RX (receive).

We can then analyze each of these activities (abort, read queue, do DMA) separately by examining the states that execute them. These states in term refer to functions (like *ok_to_read_q*) which can be found in the functions file (in alphabetical order).

We have now defined a preferred structure for a module, consisting of a state machine, encapsulating the sequential code of the module, along with a set of functions, which encapsulate the combinational code. The various states of the state machine call these functions. The state machine is the primary component of the module, and the place where we start analyzing the design. It functions like *main* in a C program. The functions are secondary components, called only by the state machine (or by each other).

Figure 3-2 shows this structure – the state machine consists of three states. State S0 calls function *foo*, and state S2 calls function *bar*.

This structure provides a systematic way of reviewing and understanding the code. In particular, it allows this analysis to be performed top-down and in sections of manageable size.

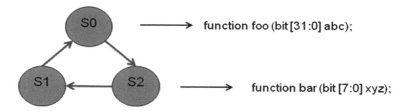

Figure 3-2 The general paradigm for state machines: states call functions.

Restructuring for a Hierarchical State Machine

Now that we have created an overall structure for the code, we can make one more improvement to the state machine: convert it to a hierarchical finite state machine (HFSM). We discuss this in the next chapter.

System Verilog

The final modification that we made to the code was to employ the SystemVerilog construct *struct*. We used this technique to encapsulate some of the common structures in the design. Consider the following example:

When the BCU prepares to do a receive (RX) DMA, it first reads four 32-bit words from the receive request FIFO. This data contains all the required information about the DMA transfer, and is composed of fourteen fields of varying width.

So as the data is popped off the FIFO word by word, the data is loaded in to the appropriate variables (registers).

In the original code, the designer thoughtfully named these with a common prefix to show that they were all associated with the receive DMA request:

```
reg            rxqreq_vld;
reg            rxqreq_ds;
reg            rxqreq_so;
reg            rxqreq_rr;
reg    [3:0]   rxqreq_tag;
reg    [12:0]  rxqreq_stripcnt;
reg            rxqreq_last_seg;
reg    [12:0]  rxqreq_len;
reg    [31:0]  rxqreq_addr;
reg    [15:0]  rxqreq_staddr;
reg            rxqreq_first_seg;
reg            rxqreq_first_dword;
reg    [12:0]  rxqreq_rbc = 0;
```

Unfortunately, there are two problems with this approach, both of which can be addressed by using the SystemVerilog *struct*.

1) There is nothing in the declaration other than the name that shows that these are all really fields in a single data structure.
2) There is nothing in the declaration that indicates where the various fields are located in the data words loaded from the FIFO.

In fact, the data words are distributed throughout the data words read from the FIFO, with a number of bits that are not used (n/u). Figure 3-3 shows the location of the fields in the four words in the FIFO.

staddr[15:0]				first_seg	first_word	rbc[12:0]		vld
n/u	ds	so	n/u[2:0]	rr	n/u[3:0]	tag[3:0]	n/u[2:0]	stripcnt[12:0]
last_seg	n/u[17:0]					len[12:0]		
addr[31:0]								

Figure 3-3 Fields in the FIFO.

The original code uses a set of macros to specify the field locations, so reading the third word looked like:

```
txqreq_last_seg[txqnum_sel] <= bcu_rdata[F_LEN];
txqreq_len[txqnum_sel] <=bcu_rdata[F_RLEN-:W_RLEN];
```

So if we blindly trusted these macros to be correct, there was no problem. But when we tried to verify that the fields were being loaded correctly, we had to refer to the functional specification (which had a table like Figure 3-3) and the macro definitions (in a separate file) and the code itself. This was more documentation than we really wanted to look at simultaneously. And it all violated the principle of locality: all the information about the data structures in the FIFO, and how they are loaded into BCU registers, should be in one place.

This problem is addressed by using a *struct*, where the code becomes:

```
struct packed {
    bit [15:0]  staddr ;
    bit         first_seg ;
    bit         first_dword ;
    bit [12:0]  rbc ;
    bit         vld;
```

(continued)

(continued)

```
//DWORD1 :
  bit  [1:0]     unused1;
  bit            ds ;
  bit            so ;
  bit  [2:0]     unused2;
  bit            rr ;
  bit  [3:0]     unused3;
  bit  [3:0]     tag ;
  bit  [2:0]     unused4;
  bit  [12:0]    stripcnt ;
//DWORD2 :
  bit            last_seg ;
  bit  [17:0]    unused5;
  bit  [12:0]    len ;
//DWORD3 :
  bit  [31:0]    addr ;
} rxqreq = 0;
```

Having defined *DWORD1* to be "63:32", we can read the entire 32 bit word with one assignment by saying:

```
rxqreq[DWORD1] <= bcu_rdata;
```

And the data from the FIFO is loaded into the correct fields in the *struct*.
Thus, using a *struct*:

1) Makes it clear that the different fields are all part of the same data structure
2) Makes it clear where the fields are located in the FIFO words
3) Reduces the lines of code necessary to read the fields from the FIFO

Using *structs*, we reduced the number of objects declared in the design (*reg's*, *wires*, and, in the new version, *structs*) from about 75 in the old version to 25 in the new version. Again, encapsulation has simplified the design.

Note: In re-coding the BCU we did not use the SystemVerilog *interface* construct, in order to remain compatible with the other modules in the design. For new SystemVerilog designs, the interface construct can be useful. See Appendix D for a discussion of the SystemVerilog *interface* construct.

Simplified Block Diagram

One of the interesting side-effects of the code restructuring is that it allows a very concise, complete and accurate drawing of the design. Figure 3-4 shows this diagram.

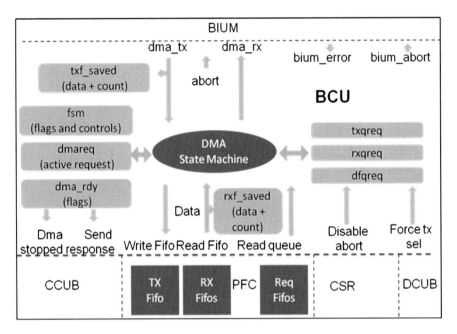

Figure 3-4 Block diagram of the BCU (center) and surrounding modules.

The design (BCU) communicates with the BIUM (bus interface), CCUB (command control), PFC (FIFO control), CSR (control and status registers), and DCUB (data control) blocks. All the possible input/output transactions are shown in the yellow arrows. The eight data structures (*fsm, dmareq, dma_redy, txqreq, rxqreq, dfqreq, rxf_save, txf_saved*) contain virtually all the signals in the design. All the sequential code is in the DMA State Machine. The only parts of the code not shown are the functions (combinational code) referenced by the state machine.

Summary

We have used the following techniques to simplify the code:

- partitioning the code into multiple files
- eliminating some syntactic fluff from the code
- restructuring the sequential code
- using functions for the combinational code
- converting the state machine to a single, sequential process
- converting the state machine to a hierarchical state machine
- using the system Verilog *struct* construct to encapsulate important structures

The result has been:

- We have reduced the state space of the design from about 2^{56} to about 2^4.
- These states are encapsulated in a hierarchical state machine (shown in the next chapter).
- We have provided a systematic encapsulation of sequential and combination code, allowing a simplified, systematic review process.
- We have reduced the number of objects declared in the design by 67%.
- We have reduced the number of lines of code by about 30%.
- We have eliminated the syntactic fluff to make the function of the code much more obvious.

Chapter 4
Hierarchical State Machines

Restructuring for a Hierarchical State Machine

In the previous chapter, we made numerous improvements in the BCU, and developed an overall structure for the code. In this chapter, we re-work the finite state machine to be a hierarchical state machine (HFSM). We start with the general model for HFSMs and then move on to re-coding the BCU state machine.

The finite state machine is central to the structure of the BCU: it defines how the design functions over time. It is like the *main* in C code: it's where we start any analysis of the code. Making the state machine as well-structured and as easy to understand as possible is critical.

General Model for HFSMs

State machines, like most aspects of design, follow the rule of seven. If there are more than about seven states in a state machine, it becomes much harder to design, to review, and to understand. The solution is to design and code it using hierarchy, and to limit the number of states at any one level of hierarchy to (at most) about seven.

Figure 4-1 shows the top level of a (generic) hierarchical state machine that consists of three states. *IDLE* is a normal state. *S1* and *S2* are composite states, indicated by the concentric circles.

Composite states are sub-state machines: state machines that are called by another state machine. Figure 4-2 shows the detail of the *S1* sub state machine.

M. Keating, *The Simple Art of SoC Design: Closing the Gap between RTL and ESL*,
DOI 10.1007/978-1-4419-8586-6_4, © Synopsys, Inc. 2011

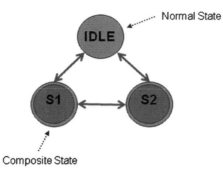

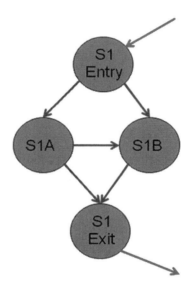

Figure 4-1 Top level state diagram for a generic hierarchical finite state machine.

Figure 4-2 Sub state finite state machine for composite state S1.

Designing and coding a state machine to be a hierarchical state machine does not change the function of the state machine at all. It is largely a notational convenience, like *functions*, that facilitate structure in our code. We code *S1* as a *task* that is called by the *case* statement in the main state machine.

There is still only one state active at any one time. For instance, if the main state is in S1, then none of the states in *S2* are active. If the main state is in *S1* and *S1* is in the *S1A* state, then the whole machine is in the *S1A* state.

Example 4-1 indicates how to code a hierarchical finite state machine, using tasks. The full code for this example is in Appendix B. Figure 4-3 shows the top level of the state machine. This simple state machine reads a 32-bit packet and then sends it out as two 16-bit words.

```verilog
enum {IDLE, GET_PKT, SEND_PKT} tx_state ;
enum {GP_READ, GP_DONE } get_pkt_state ;
enum {SP_DEST, SP_PAYLOAD, SP_DONE} send_pkt_state;

// ------------ main state machine ----------------

always @ (posedge clk or negedge resetn) begin
 case (tx_state)
  IDLE : if (pkt_avail) tx_state <= GET_PKT ;
  GET_PKT : begin
   get_pkt () ;
   if (get_pkt_state == GP_DONE)
      tx_state <= SEND_PKT ;
  end
  SEND_PKT : begin
   send_pkt () ;
   if (send_pkt_state == SP_DONE)
      tx_state <= IDLE ;
  end
 endcase
end

// ------------ get_pkt sub state machine ----------

task get_pkt () ;
 case (get_pkt_state)
  GP_READ : begin
   input_packet.dest <= data_in[31:16] ;
      ...
   get_pkt_state <= GP_DONE ;
  end
  GP_DONE : begin
   get_pkt_state <= GP_READ ;
  end
 endcase
endtask

// ------------ send_pkt sub state machine -------

task send_pkt() ;
 case (send_pkt_state)
      ...
 endcase
endtask

endmodule
```

Example 4-1

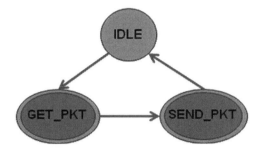

Figure 4-3 Top level state machine diagram for Example 4-1.

In the classic paradigm for hierarchical state machines [10], there is only one entry state into a sub-state machine. Referring back to Figure 4-2, all entries into the sub-state machine must start at state *S1_ENTRY*. We are not allowed to jump directly from *IDLE* (in the top state machine) to *S1A*. Similarly, there is only one exit state from the S1 sub-state machine, *S1_EXIT*. We are not allowed to jump from an intermediate state (like *S1B*) back to *IDLE* in the main state machine.

In Example 4-1, we enforce these rules by the way we code the state machine. The sub-state machines (*get_pkt* and *send_pkt*) are coded as tasks, which are called by the main state machine. What happens in the sub-state machine – specifically what happens on entry and on exit - is completely determined by the task.

These rules place some limitations on the state machine design, but more than compensates by providing a clean structure. Such a clean structure is essential when designing large, complex state machines.

The most significant restriction imposed by this state machine coding style is the fact that it is purely sequential. In some designs, there are primary outputs that are combinational, not registered, and that need to be driven as soon as a state machine enters a particular state.

For instance, we can imagine in Example 4-1 that we might need to drive some output (let's call it *comb_out*) as soon as we enter the *GP_READ* state in the *get_pkt* sub-state machine. There is not an elegant way to do this in SystemVerilog. (Later in the book we propose some extensions to SystemVerilog to fix this.)

The best solution is to have a separate combinational process, something like:

```
always_comb
if ((tx_state      == GET_PKT ) &&
    (get_pkt_state == GP_READ))
comb_out = 1;
```

Converting the BCU to a HFSM

We now turn to the challenge of converting the BCU state machine to a hierarchical state machine. The biggest challenges have to do with separating the states into clean sub-state machines with one entry state and one exit state. The original BCU state machine has a rather spaghetti-like graph, and untangling it was our biggest challenge.

Figure 4-4 shows the complete state machine state diagram.

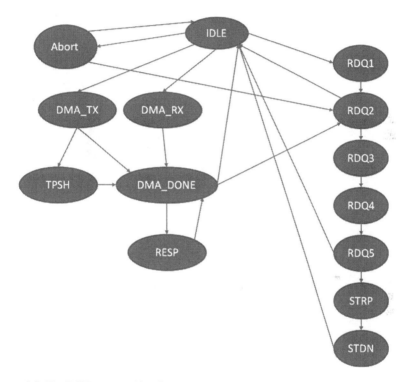

Figure 4-4 The BCU state machine diagram.

The state machine appears to consist of two simple states (*IDLE* and *ABORT*) and two candidate composite states: *ReadQ* (which fetches the DMA request), and the DMA process itself. So our goal is shown in Figure 4-5.

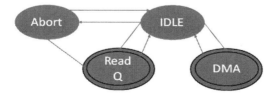

Figure 4-5 Top level of the hierarchical version of the BCU state machine.

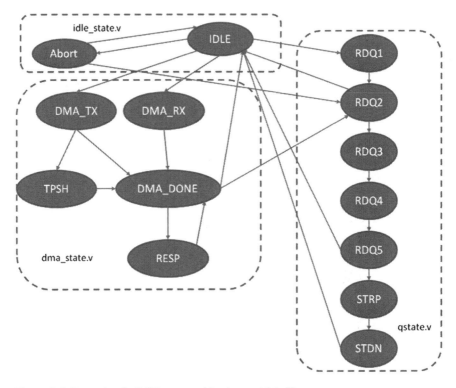

Figure 4-6 Separating the BCU state machine into multiple files.

To make analysis easier, we partitioned the state machine code into three separate files as shown in Figure 4-6. We moved the case statements for the different DMA states into *dma_state.v* and moved the *case* statement for the read queue states into *qstate.v*. We used `include` to insert the code in the right place in the top file, so we could test our changes every step of the way.

Our job now was to re-work the state machine so that we could convert *qstate.v* into a single task and convert *dma_state.v* into a single task. Then we would have a true hierarchical state machine.

We observed that *dma_state.v* has two entry points: *DMA_TX* and *DMA_RX*. So we decided to add an explicit *DMA_ENTRY* state to the DMA sub-state machine.

We also noted that *DMA_DONE* is already a single exit point for *dma_state.v*, but this state is very complex and interacts with all the other states. So we tentatively added another state, *DMA_EXIT*, which now just deals with setting the *done* bit and exiting cleanly. This minimized the changes need to the *DMA_DONE* case statement.

The DMA state is now a sub-state machine, as shown in Figure 4-7.

We observed that qstate.v has two entry points (*RDQ1* and *RDQ2*) and two exit points (*RDQ5* and *STDN*). In this case we definitely needed to add separate *RDQ_ENTRY* and *RDQ_EXIT* states to meet the single-entry, single-exit design goal. Figure 4-8 shows the results.

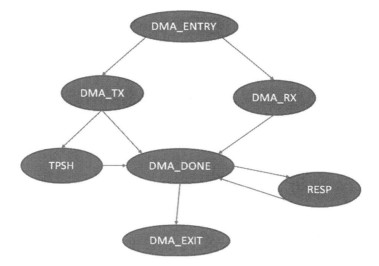

Figure 4-7   The DMA sub state machine.

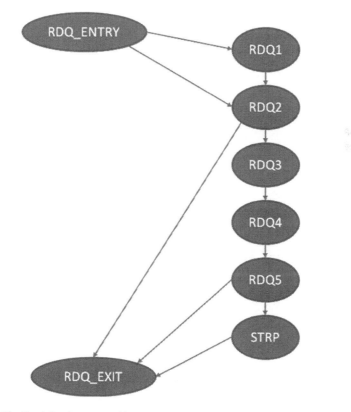

Figure 4-8   The Read Q sub state machine.

Adding the explicit entry and exit states to the DMA and RDQ state machines did not come for free – it added two clock cycles to each DMA and to each queue read. But analysis and extensive testing showed that these extra cycles did not adversely affect performance. The modified design passed all regression tests.

If we had to, we could do a more extensive re-design that would have eliminated these extra cycles. But it would have made the code more complex and the rework effort much greater.

We have done some additional experiments where we designed and coded a state machine from the outset to be hierarchical. In all these cases we were able to avoid adding any extra cycles.

Our first conclusion from this experience was that hierarchical state machines, done properly and architected from the outset during the specification process, can be implemented in SystemVerilog with no performance penalty.

Our second conclusion was that design reviews of hierarchical state machines are much more successful than those of flat state machines. We participated in design reviews of this hierarchical state machine, as well as other hierarchical and flat state machines used in other projects. Consistently, after design reviews of modules that used HFSMs, the reviewers said that they understood the design completely and were confident that it was correct. Equally consistently, reviewers of (complex) flat state machines said that they understood some, but not all, of the behavior of the design and were not nearly as confident that it was correct.

Our third conclusion was that, although we can design HFSMs in SystemVerilog, language restrictions still prevented us from coding the state machines as we would really like. In particular, we find the generation of combinational outputs from the state machine very awkward. The result is that we have developed some proposals for adding a (hierarchical) state machine primitive to SystemVerilog. We discuss these proposals in Chapter 11.

Chapter 5
Measuring and Minimizing State Space

State space is a key metric for the complexity of a digital design. For most designs, it is the most important metric: the size of the control state space is the best indication of the complexity of the design. If this state space is very large, it is more likely to have bugs, to have bugs that are hard to detect, and to have bugs that are hard to fix without injecting other bugs in the process.

In some designs, the data path (the combinational functions) can be quite complicated. Video and DSP algorithms can involve arithmetic operations on matrices, with nested loops of multipliers. But even in these designs, the really difficult bugs are most often found in the control logic.

This chapter focuses on analyzing the state space of RTL designs. Our first objective is to measure the state space. Then we discuss how to minimize state. But our most important objective is to make the entire state space of a design easy to understand. This will allow us to get it right when we design it; it will allow us to verify it, to fix it and to extend it as necessary.

To make the state space as easy to understand as possible, we will make it:

- explicit
- structured
- minimal, in terms of the total number of states

Input, Output, and Internal State Space

There are three state spaces that contribute to the complexity of the overall design:

- input state space
- output state space
- internal state space

M. Keating, *The Simple Art of SoC Design: Closing the Gap between RTL and ESL*, DOI 10.1007/978-1-4419-8586-6_5, © Synopsys, Inc. 2011

The input state space of a design consists of all the different states that its input signals can represent. In its most naïve form, the size of the input state space is 2^N, where N is the number of input wires to the module.

The output state space of a design consists of all the different states that its output signals can represent. In its most naïve form, its size is 2^M, where M is the number of output wires from the module.

The internal state space of a design consists of all the different states that its internal registers can represent. In its most naïve form, its size is 2^P, where P is the number of register bits inside the module.

Since the output state space of one module is the input state space of other modules, we will focus primarily on the input state space and internal state space in the rest of our analysis.

Preliminary Calculations of State Space

Using our naïve measurements of state space, we can calculate the input, output and internal state space from the interface and signal declaration parts of the RTL code. That is, given the following code (a small subset of the declarations for the BCU example):

```
module bcu (
   input clk,
   input hreset,

   output              idle,
   output              fifo_busy,
   output   reg        dma_req,
   output   reg  [31:0] dma_addr,
   output   reg  [12:0] dma_count,

   input  [31:0] rdata,
   input         dma_done,
   input         dma_error,
   input         wdata_pop,
   input         rdata_push
);

reg [3:0]   state_bcu;
reg [1:0]   fsm_q_sel;
reg         fsm_dmareq_vld_set;
reg [12:0]  dmareq_len;
reg [31:0]  dmareq_addr;
```

Example 5-1

We can calculate the output state space = 2^M, with M the number of outputs :

Outputs: *idle, fifo_busy, dma_req, dma_addr* (32 bits), *dma_count*(13 bits)
M = 1 + 1 + 1 + 32 + 13 = 48
Output State Space = 2^{48} = 281 trillion

We calculate the input state space = 2^N:

Inputs: *rdata*(32 bits), *dma_done, dma_error, wdata_pop, rdata_push*
N = 32 + 1 + 1 + 1 + 1 = 36 (not counting clock and reset)
Input State Space = 2^{36} = 68 billion

We calculate the internal state space = 2^P:

Registers: *state_bcu* (4 bits), *fsm_q_sel* (2 bits), *fsm_dmareq_vld_set, dmareq_len* (13 bits), *dmareq_addr* (32 bits)

P = 4 + 2 + 1 + 13 + 32 = 52
Internal State Space = 2^{52} = 4 $*10^{15}$

These state spaces are huge; even the smallest is way too large to enumerate explicitly, or draw on a state diagram. But we notice an important fact about the output state space: *dma_addr* and *dma_count*, by their names, imply that they are data, not control signals.

In this particular design, the module sets up the DMA; it is given the DMA address and the DMA count as inputs (derived from *rdata*). It just passes this address and count on to its outputs. These signals do not affect the decisions made by the module in setting up the DMA. The value of the DMA address, for example, has no effect on whether or how the DMA is performed. Note that we know this information only by inspection of the RTL code of the module. There is no information in the input/output declarations that identify which inputs/outputs are data and which are control.

Armed with this information, we can recalculate the output state space:

Outputs: *idle, fifo_busy, dma_req,*
M = 1 + 1 + 1 = 3
Output State Space = 2^3 = 8

Similarly, with the input state space, we review the RTL and learn that rdata is, in fact, data and not control; it does not affect the actual input state space. We recalculate:

Inputs: *dma_done, dma_error, wdata_pop, rdata_push*
N = 1 + 1 + 1 + 1 = 4 (not counting clock and reset)
Input State Space = 2^4 = 16

These are dramatically smaller numbers; we can, in fact, write down a list of all possible states, or draw this state space in the form of a diagram. This capability indicates that these are state spaces we are likely to be able to understand completely.

Shallow vs. Deep State Space

Just as we distinguish between data and control for input and outputs state space, we distinguish between shallow and deep state space in analyzing the internal state space.

Deep state space consists of counters, memory, and similar elements: states that we would never draw as distinct states in a state machine diagram. Shallow state space consists of the states we normally think of as the state space in a design: the state space described by a state machine.

Consider the following trivial state machine:

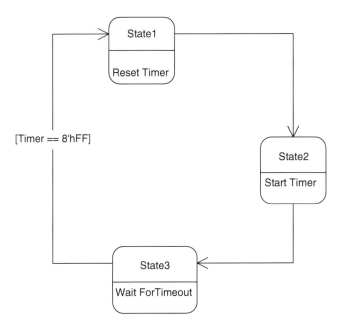

Example 5-2

This simple design resets the timer, starts the timer and then waits for it to count to 255 before resetting. The eight bit timer has 256 states, but these are considered deep states – states we would never draw in a state diagram. The state machine itself consists of three (shallow) states – *State1*, *State2* and *State3*.

Note that there is a close association between timer states and the shallow states of the state machine. *State1* sets the timer to zero. *State2* starts the timer counting. In *State3*, the timer is counting. As far as a state machine is concerned, the timer only has three "interesting" states: zero, counting, and 8'hFF.

This distinction between deep and shallow state space is important in correctly calculating the complexity of a design. The above state machine would be no more complicated if it had a 12 bit counter – it would just wait in *State3* longer. Similarly, reducing the counter to 4 bits would not make the design simpler.

Returning then to Example 5-1, given the code:

```
reg [3:0]   state_bcu;
reg [1:0]   fsm_q_sel;
reg         fsm_dmareq_vld_set;
reg [12:0]  dmareq_len;
reg [31:0]  dmareq_addr;
```

We calculated the internal state space as:

$P = 4 + 2 + 1 + 13 + 32 = 52$
Internal State Space $= 2^{52} = 4 * 10^{15}$

We observe (by analysis of the RTL) that *dmareq_len* is just temporary storage for the DMA count. The actions of the state machine do not depend on the particular value of *dmareq_len*. It is loaded with a count value (from the module inputs) and then decremented as the DMA progresses. The state machine then moves to the next state when *dmareq_len* is zero. It is just like the counter in Example 5-2. That is, it is deep state.

Similarly, analysis shows that *dmareq_addr* (the DMA address) and *fsm_q_sel* (which DMA source to use) are deep state.

We now recalculate the internal state space:

```
reg [3:0]   state_bcu;
reg         fsm_dmareq_vld_set;
```

$P = 4 + 1 = 5$
Internal State Space $= 2^5 = 32$

This space is much more understandable than 2^{52}, and it is something we could draw in a state diagram.

Further analysis shows that *state_bcu* (as its name implies) is the state variable for the state machine. So it is clearly shallow space, and thus part of the internal state space calculation.

The signal *fsm_damreq_vld_set* is more complicated. It is a flag set in one of the states of the state machine. Since its value does directly affect state (it determines which state the state machine goes to next), we consider it shallow state. In general, any variable that is not obviously deep state, we consider part of the shallow state space.

The Cross Product of State Spaces

In the previous example, we considered a design with only one state machine, and with only one state variable. If instead, we had seen the following code:

```
reg [3:0]    state_bcu;
reg [2:0]    state_foo;
reg          fsm_dmareq_vld_set;
```

And if our analysis showed that both *state_bcu* and *state_foo* were state variables for state machines, then our calculation of internal state space would be:

$P = 4 + 3 + 1 = 8$
Internal State Space $= 2^8 = 256$

Note that the addition of a very small (eight state) state machine has grown the internal state space from one we could draw (and understand) quite easily, to one that would be quite difficult do draw or understand!

Here is another way of looking at this issue:

If there is more than one state machine in the module, then the internal state space of the module is equal to the cross product of the state spaces of the state machines. Consider Example 5-3:

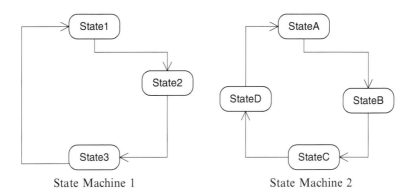

State Machine 1   State Machine 2

Example 5-3

Here the total state space is the product of the number of states in the first state machine times the number of states in the second state machine, that is

Total State Space $= 3 * 4 = 12$.

If the two state machines are genuinely independent, then this is the correct calculation. That is, when the first state machine is in *State1* the second state machine can be in any one of its four states, and when the first state machine is in *State2*, the second state machine can be in any of its four states, and when the first state machine is in *State3*, the second state machine can be in any one of its four states. This is what we mean by fully independent state machines.

But in most designs, there is some correlation between the two state machines. For instance, when the first state machine is in *State1*, there may be a restriction that the second state machine can only be in *StateA* or *StateB*. In this case, the intended state space is smaller than the cross product of the two state machines. But it may be nontrivial to determine by inspection what this restricted, intended state space is. In many cases, the original intent of the designer may be impossible to determine from the design itself.

For instance, there may be nothing in the code that logically restricts the second state machine from being in *State D* when the first state machines in *State1*. It may be that the expected input transactions can never produce this combination of states. But this expectation about sequences of inputs is not explicit in the design of the state machines.

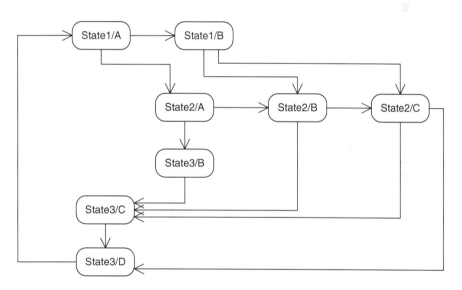

Example 5-4

However, if the code is re-written to have only one combined state machine, the rules become explicit about which (combined) states are possible. Let us suppose that the intended behavior of the (combined) state machine is as shown in Example 5-4.

Note that in the Example 5-3, there were seven states drawn in the two state machines, but there were 12 possible states for the combination of the two (the cross product 3 * 4 = 12). In our new design, there are eight states drawn and eight possible states. That is, the state space now is completely explicit – the drawing (and code) show exactly which states are possible and which are not. No information is left in the designer's head (about legal vs. illegal states). It is all there on paper for everyone to see.

Our code-based calculation changes as well. When we had two state machines, one with three states and one with four states, our code might look like:

```
reg [1:0]    state_123;
reg [1:0]    state_abcd;
```

Then our initial calculation would be:

P = 2 + 2 = 4
Internal State Space = 2^4 = 16

But we know from the design that this is wrong! We only have 12 states. But because we used two bits as the state variable for the three-state machine, our code implies that the space is larger than it really is. This is where SystemVerilog's enumerated type is so valuable.

Instead we code Example 5-3 as:

```
enum bit {State1, State2, State3} state_123;
enum bit {StateA, StateB, StateC, StateD} state_abcd;
```

Now we are back to the 12 states we expected. Our internal state space calculation is now:

Internal State Space = 3 * 4 = 12

That is, the cross product of the two enumerated state variables.
In our improved design (Example 5-4), with one state machine, we have:

```
enum bit   {State1_A, State1_B, State2_A, State2_B,
            State2_C, State3_B, State3_C, State3_D}
            state_123;
```

This simple exercise in counting the state space shows the great advantage of having just one state machine in any one module and using enumerated types to declare the state variable.

Sequential Processes and Internal State

Even when a module has only one (explicit) state machine, the effective shallow state of the module may be much larger than that of the state machine.

Any time we code a sequential process, we are creating state. That is, anytime we create a flip-flop by entering code such as:

```
bit a;
always_ff @(posedge clk) a <= b;
```

This code creates a one bit state machine consisting of two states; the variable *a* can have the value zero or one, depending on the value of *b*. In the context of the module, this may be deep or shallow state, but unless we have very explicit indications that this is deep state, we have to treat this as shallow state.

By adding a sequential process such as the one above, we clearly add to the total state space the design. For instance, if the above piece of code is included in the module that has State Machine 1 and State Machine 2 from Example 5-3, then the state space calculation becomes:

Total State Space = 3 * 4 * 2 = 24.

Thus, the addition of a single flip-flop has doubled the state space for the module.

There is no faster way to cause the state space of a module to explode than to have a large number of independent sequential processes.

Encapsulating Sequential Code

Encapsulation, abstraction, and information hiding are the key tools for managing complexity in a design. In one form or another, these all involve the basic strategy of divide and conquer.

In (System)Verilog, the *function* provides a way of encapsulating combinational code. There is no equivalent of encapsulation method for sequential code. There is no way to place an *always@(posedge clock)* or *always_ff* inside a structure that will hide local values and calculations, and which will make explicit what global signals the process depends on.

The state machine is the closest mechanism we have for encapsulating sequential code. With a state machine we can at least add structure to sequential code, making the state space explicit. We can make the state variable explicit, and, using enumerated types, we can make the legal values explicit as well.

The great advantage of the state machine is that it makes the time evolution of the circuit explicit. If we have a large number of sequential processes scattered throughout the code for a module, then we have to analyze the behavior of each of these processes for every cycle. This can become extremely complex.

On the other hand, when analyzing a state machine we can analyze the behavior of the code state-by-state. Thus, by using a state machine, we can change a global problem (analyzing all the sequential processes in the module simultaneously for each cycle) to a series of local problems (analyzing each state independently).

State Machines as Sequential Processes

To take advantage of the state machine as an encapsulation device, of course, it must be coded as a sequential process. The other standard coding style, using two processes, results in code like Example 5-5:

```
case (state)
  STATE1: begin
    if (condition1) begin
      next_state = STATE2;
      foo = bar;
    end else begin
      foo = bar;
    end
  end
  .
  .
  .
endcase

always_ff (posedge clk) foo_reg <= foo;
```

Example 5-5

That is, we end up with lots of sequential processes outside the structure of a state machine – and the state space loses its structure and starts to explode in size.

Hierarchical State Machines

As mentioned in the previous chapter, we have found that designing and analyzing a state machine with many states to be extremely difficult. Hierarchical state machines address this problem, and are an essential tool for designing complex systems.

In the case of the BCU, the state machine was not really that complex. So in this section we give an example of a much more complex state machine. This state machine controls the protocol layer of the USB OTG core. The original design consisted of a very large, unstructured state machine consisting of 10's of pages of code. We restructured the code to be a hierarchical state machine.

Examples

The following diagrams show three pages of a state chart drawing for a hierarchical state machine. Figure 5-1 shows the top level of the state machine.

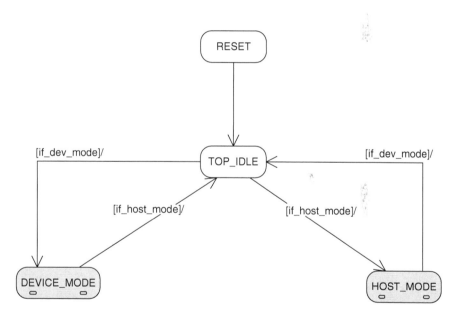

Figure 5-1 Top level USB OTG hierarchical finite state machine.

At the top level, state machine has two composite states (states that call sub state machines): device mode, and host mode. Figure 5-2 shows the host mode sub state machine. This sub state machine consists of five states plus two composite states: host out and host in.

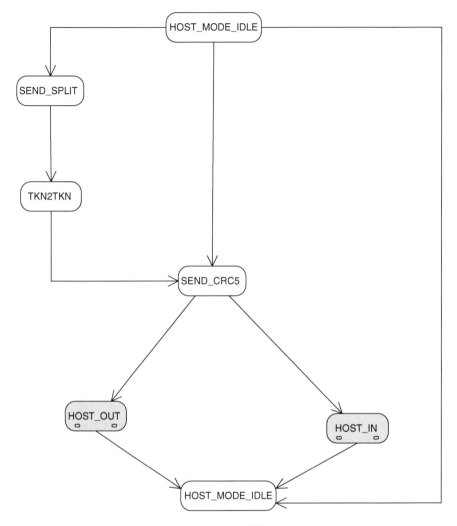

Figure 5-2 Host Mode sub state machine for USB OTG.

Figure 5-3 shows the detail for the host out sub state machine. This is a leaf level sub state machine, consisting of six states and no composite states.

These are three of the eight drawings that comprise the complete state machine. The entire state machine consists of about 30 states. Thirty states are too many to

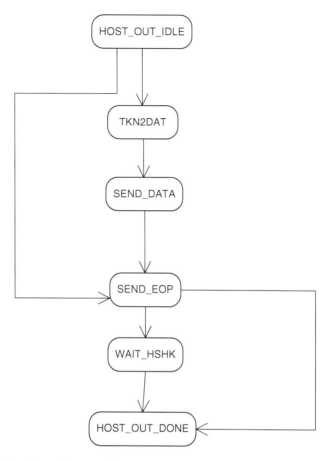

Figure 5-3 Host Out sub state machine.

draw on a single piece of paper, and way too many to analyze as a flat design. But using a hierarchical state machine, we can limit the number of states in any one drawing to a maximum of eight states. We can analyze each sub-state machine, each drawing, as a separate object. This use of hierarchy greatly simplifies both design and analysis of the state machine.

Again, the BCU example shown earlier, with a total of fourteen states, is simple enough that going to a hierarchical state machine gives only a small advantage. But for a complex state machine like the USB, with thirty states, hierarchy is essential to keep design complexity manageable.

Summary- Counting State

Input State Space

1. Examine the input declarations. Remove the ones that are data – that is, ones whose values do not affect the operation of the design.
2. For each remaining input variable, determine state space of that variable:

   a. If it is an enumerated type, take the number of values
   b. If it is not an enumerated type, take 2^N, where N is the number of bits in the input variable

3. Multiply the state space sizes of all the input (control) variables to get the full input control state space.

Output State Space

1. Examine the output declarations. Remove the ones that are data – that is, ones whose values are not affected by the operation of the control path.
2. For each remaining output variable, determine state space of that variable:

   a. If it is an enumerated type, take the number of values
   b. If it is not an enumerated type, take 2^N, where N is the number of bits in the output variable

3. Multiply the state space sizes of all the output (control) variables to get the full output control state space.

Internal State Space

If there is more than one sequential process in the module:

1. Examine the declarations of the variables that are assigned a value in sequential processes

   a. Remove the ones that are data – that is, ones whose values do not affect the operation of the design.
   b. Remove the ones that are deep state – counters, memory, and so on.

2. For each remaining variable, determine state space of that variable:

   a. If it is an enumerated type, take the number of values
   b. If it is not an enumerated type, take 2^N, where N is the number of bits in the variable

3. Multiply the state space sizes of all the internal (control) variables to get the full internal (shallow) control state space.

If all the sequential code is in one finite state machine, then the internal state space is the number of states in the state machine.

State Space for Hierarchical State Machines

The above analysis of internal state space does not distinguish between flat and hierarchical state machines. We would like a metric that rewards hierarchy and reflects how it helps reduce the effective complexity of a design.

We consider such a metric to be an open question. But we suggest that something like the following would be reasonable:

1. Count the total number of sub-state machines in the design – call this SS
2. Find the (sub-)state machine with the largest number of simple states – call the number of simple states in this (sub-)state machine M
3. Complexity = SS + M

For a flat 32-state machine:

SS = 0
M = 32

Complexity = SS + M = 32.

For the same state machine coded as a hierarchical state machine, with a total of four sub-state machines, each with eight simple states:

SS = 4
M = 8

Complexity = 4+8 = 12.

So the hierarchical design is judged to be roughly three times less complex than the flat design.

Chapter 6
Verification

This chapter discusses the challenges and opportunities of verifying RTL designs. In particular, we explore the opportunities presented by our proposed design approach. This approach – encapsulating the sequential code in a single state machine and combinational code in a set of functions – allows us to develop some powerful module level verification techniques. These techniques get us closer to a complete verification strategy than is possible using traditional design and coding practices.

In general, verification is an unbounded, and often ambiguous, problem. The one well-defined, bounded problem in verification is whether we have exercised all the behaviors of the circuit. That is, can we define a set of stimuli that, applied to the design under test, will exercise all the functionality of the design. This will be our focus for this chapter.

In developing this verification strategy, we will need to define more formally some concepts introduced earlier:

- Data vs. Control inputs and outputs
- Shallow vs. Deep internal states

First, though, we make some observations about verification; these observations will help develop a sense of what can and cannot be achieved through (simulation-based) verification. We focus on trying to answer the question: "When are we done simulating?"

M. Keating, *The Simple Art of SoC Design: Closing the Gap between RTL and ESL*, DOI 10.1007/978-1-4419-8586-6_6, © Synopsys, Inc. 2011

Some Simple Examples of Verifiable Designs

> If I want to test a light switch in my house, I flip it on and then I flip it off. I'm done. It's tested. Additional flipping of the switch will give me no more information about the switching circuit.
>
> Note: There are lots of other things I might want to know – like the lifetime of the switch, whether it works in all weather, and so on. And these might be discovered by additional testing. But for the purposes of this discussion, I am only interested in the functionality of the switch. Once I have flipped it on and off, I know everything about the function of the switch. In particular, I will know whether the up position is on or off. If I find out that the up position is off, then I may decide the switch is backwards – functionally incorrect. But the key here is that I know when to stop testing. How I interpret the results of the test is outside the scope of this discussion. We are focused exclusively on the question: When am I done testing.

If we want to test even a moderately complex piece of RTL code, knowing when we are done is a daunting task. Most verification teams do not even attempt to achieve complete verification, and instead test as thoroughly as they can in a reasonable amount of time. Then they ship it, and hope that it doesn't come back.

This is a sorry state of affairs. In all other aspects of chip design, we know when we are done and are ready for tape-out. Static timing analysis, DRC and extraction, and power analysis all have well-defined pass/fail criteria for completion. But verification does not.

So let's investigate what it would mean if we could completely verify an RTL design.

We will start with an AND gate.

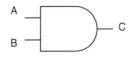

Clearly we can completely verify this circuit. In simulation, we simply assert the four possible input patterns to A and B and check the output at C. Once we have done that, additional simulation will give us no new information on the behavior of the circuit.

Thus, for a relatively simple combinational logic, we can completely test the circuit. On the other hand, a 32 x 32 multiplier requires 2^{64} input patterns to be applied. Even with a superfast simulator, where we can apply an input vector every nanosecond, it would take more than 500 years to completely verify this design. Even though we can define a complete set of tests for the multiplier, it is not practical to exercise them all.

Next, let's consider a simple flip flop with an AND gate:

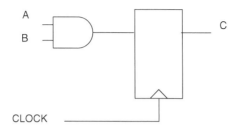

Again we can completely verify this design. In simulation, we assert the same four input patterns as we did for the AND gate, and after each pattern we toggle the clock. We're done. It's tested.

That is, we have simulated all possible input, output, and internal states for the design.

On the other hand, consider a design with 64 flip-flops, each of which is completely independent from the other 63. From Chapter 5, we know that the state space of this design is 2^{64}. So we are dealing with the same number of tests as in the case of the 32 x 32 multiplier. And again it will take 2^{64} input patterns, and again we're looking at more than 500 years to cover the complete state space.

Clearly there is some dividing line between a simple circuit with one gate and one flip-flop and the kind of complex designs that end up in an SoC. The question is: where is this dividing line, and are there tricks we can use to bring a design back over the line from being unverifiable to being verifiable?

Verification Overview

Ultimately, both verification and design are human activities. In design, a human being converts some form of specification into a detailed design. No matter how powerful the design tools or methodology, the key actor in design is the human designer. The basic intent of the design has to be defined by a person who understands the target application, the problem it is trying to solve.

In verification, we check this person's work. Another set of eyes addresses the same task, from a different perspective. The RTL designer describes the detailed behavior of the design from the inside out. The verification engineer defines the behavior of the design from the outside in, by writing a test bench that models the environment that the design will go into. But ultimately, it is just a case of two different people looking at the same specification and developing two representations of that specification and comparing them.

Of course, each representation has bugs in it. Typically, verification engineers spend at least as much time debugging the test bench as they do debugging the design itself.

One approach to improving verification is to add additional representations of the specification. For instance, it is possible to develop a set of properties for the design and to prove formally that the design meets these properties. In addition, it is possible to insert a large number of assertions in the design, which essentially represent the intended behavior of the design in yet another format.

Each of these approaches makes sense, since they add additional pairs of eyes to look at the design and the design problem. Ultimately, each also involves additional debug, additional tools, and additional effort.

In this chapter we focus primarily on two pairs of eyes: the designer and the verification engineer. We also look at how the two can work together to produce a more verifiable design and a more complete test bench.

Goals of Complete Verification

The initial goal of complete verification is to exercise completely all the functionality that is actually in the code. Missing functionality is an important problem, but it is essentially an unbounded problem. Similarly, whether the functionality of the design is the desired functionality is an ill-defined problem. This desired functionality is typically captured in a natural language specification that is incomplete and ambiguous.

But, as mentioned earlier, the one well-defined, bounded problem in verification is whether we have exercised all the behaviors of the circuit. That is, can we define a set of stimuli that, applied to the design under test, will exercise all the functionality of the design.

The functionality of the actual code is bounded, even if it is huge. So this goal of completely exercising it is, at least in theory, achievable.

To support the goal of complete verification – in this sense of a complete set of stimuli - we have to be able to measure the completeness of the verification suite. One technique is to write functional coverage objects and use them to determine what functions of the design have been exercised. The problem with this approach is that it introduces yet another design representation. In addition to the RTL and testbench, we need to write (and debug) a complex set of assertions (coverage objects).

Instead, we focus on how we can extend code coverage to give us the completeness metric we need.

Verifying State Machines

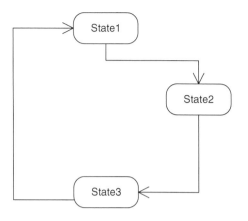

Example 6-1

For the state machine in Example 6-1, verification is pretty straightforward. Clock the state machine through all three states, back to the original state, and we are done. In particular, if the code looks like this:

```
case (state)
  State1: state <= State2;
  State2: state <= State3;
  State3: state <= State1;
endcase
```

Then once code coverage tools report that each line of code has been executed during simulation, we have achieved complete verification in the following sense:

- The entire state space has been covered; during simulation, the circuit has been put in every state, and every possible transition from one state to another has occurred
- Additional simulation will not expose any behaviors not already exposed

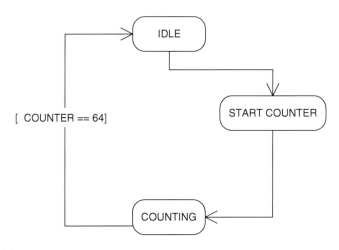

Example 6-2

In Example 6-2, things are slightly more complicated. In addition to the three shallow states, there is a counter that provides deep states.

```
case (state)
  IDLE: state <= START_COUNTER;
  START_COUNTER: begin
    counter <= 0;
    state <= COUNTING;
  end
  COUNTING: begin
    counter <= counter + 1;
    if (counter == 64) state <= IDLE;
  end
endcase
```

If the line coverage tells us that we have executed every line of code, then we know we have covered the shallow state space. But what about covering the deep state space – the counter?

We observe the following about the deep state space:

1. The counter must be at least 7 bits, but many of its values (65 to 127) are unreachable. A well-designed state machine has no unreachable shallow state, but deep state variables (like counters) may have many unreachable states.
2. The counter is set to a specific value and then modified by the state machine (the statements that clear and then increment the counter). This is what happens with

deep state – it is set to a value by the state machine (shallow state space) or by an input (input state space).

3. The counter value is used to modify the state space of the design. When the counter reaches 64, the shallow state is changed (it goes back to IDLE). Again, this is what deep state is used for – it eventually appears in an "if" (or "case") statement to decide whether and how to modify the state of the module.

4. There are many equivalent states for the state machine. While it is in the COUNTING state, nothing in the circuit changes except the counter value. If we look at the state space as an ordered pair (shallow state, deep state), the states (COUNTING, counter == 1) and (COUNTING, counter == 2) are identical as far as the rest of the circuit – and the outside world – are concerned.

These observations affect how we verify – and how we measure completeness of verification – when we consider deep state space. We note that simple line coverage does not tell us whether we have ever reached the state (COUNTING, counter == 64), and hence returned to the IDLE state. Line coverage only tells us that we executed the statement

```
if (counter == 64) state <= IDLE;
```

Conditional coverage tells us whether we have executed this line with all possible values of the conditional expression – that is, whether we have executed it with "counter == 64" true and with "counter == 64" false. Once we know that we have executed this line in both conditions, we know:

1. We have tested the reachable states of the deep state variable "counter" – the state where it is 64, and the state where it is less than 64 (the 64 "equivalent" states where it is 0-63).

2. We have executed the transition from the COUNTING state back to IDLE.

As a result, we know that we have completely exercised the circuit, and additional simulation will not expose any new behavior.

So with 100% line and conditional coverage, we've achieved complete verification – in our restricted sense of completely exercising the circuit.

Example: The BCU

We now consider a more realistic state machine. Below is code taken from the DMA controller (BCU) example described in Chapter 3.

One state in the state machine is RDQ1:

```
RDQ1: begin
  case (fsm.q_sel)
    TXQ: begin
        txqreq[fsm.txqnum_sel].first_seg <= 1'b1;
        if (!fifo_busy && !q_empty(fsm.q_sel))
            state_bcu <= RDQ2_TXQ;
      end
    DFQ: if (!fifo_busy && !q_empty(fsm.q_sel))
            state_bcu <= RDQ2_DFQ;
  endcase
```

Where:

- *fsm.q_sel* is deep state (a register holding information on which DMA queue is active)
- *txqreq[fsm.txqnum_sel].first_seg* is also a deep state register
- *fifo_busy* is an input to the module
- *q_empty* is a function

We have already shown that line and condition coverage can tell us whether we have completely exercised the design when we have deep state and shallow state only. But what about functions? The function *q_empty* is shown below:

```
function q_empty (input fsm_q_sel_type fsm_q_sel);
  case (fsm_q_sel)
    TXQ: q_empty = bcu_txq_empty[fsm.txqnum_sel];
    DWQ: q_empty = bcu_dwq_empty;
    DFQ: q_empty = bcu_dfq_empty;
    RXQ: q_empty = bcu_rxq_empty;
  endcase
endfunction
```

Where:

- *bcu_txq_empty, bcu_dwq_empty, bcu_dfq_empty, bcu_rxq_empty* are all inputs to the module
- *fsm.txqnum_sel* is a deep state register in the module.

Thus, the function is just a convenient way of describing (encapsulating) a complex function of inputs and deep state. When we use functions to encapsulate combinational code, all we are doing is encapsulating a (sometimes complex) combination of deep state and input state. When the state machine code says:

```
if (!fifo_busy && !q_empty(fsm.q_sel))
    state_bcu <= RDQ2_TXQ;
```

then the function *q_empty* is just an abbreviation for a more complex conditional statement. If we were to in-line the equivalent code, it would look like:

```
if (!fifo_busy && !(
    ((fsm_q_sel == TXQ)&& bcu_txq_empty
      [fsm.txqnum_sel])    ||
    ((fsm_q_sel == DWQ)&& bcu_dwq_empty)    ||
    ((fsm_q_sel == DFQ)&& bcu_dfq_empty)    ||
    ((fsm_q_sel == RXQ)&& bcu_rxq_empty)    ||
)
    state_bcu <= RDQ2_TXQ;
```

Thus, all of our analysis - that conditional and line coverage measure the completeness of verification – is still valid *except* for the fact that the function is called more than once.

If the function were only called once in the whole design, then code coverage of the function plus code coverage of the state machine would tell us everything about the completeness of coverage. But the fact that the function *q_empty* is called twice leaves some ambiguity – we don't know that the function was completely exercised during each call.

On the other hand, if we have a tool or script that rewrites the code, in-lining the functions in the state machine, then coverage (of the rewritten code) would give us the desired information about the completeness of verification.

Thus we now have a general model of a state machine:

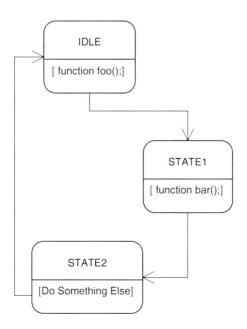

The state machine has a set of states and a set of functions that are called by states.

Our basic argument is this: in a state machine such as the one above, code coverage and complete verification are equivalent. That is, if we can accomplish complete line and conditional coverage, we have completely exercised the functionality of the code.

A Canonical Design

Based on our work so far, we can now construct a standard or canonical representation of a module. For a module that is coded in this manner, we can use (an extended version of) code coverage to provide us with a useful metric for how much of the function of the module has been exercised during simulation.

Figure 6-1 is a diagram of a module in canonical form. If the module has been designed in the style described in Chapter 3, then this is exactly what it looks like.

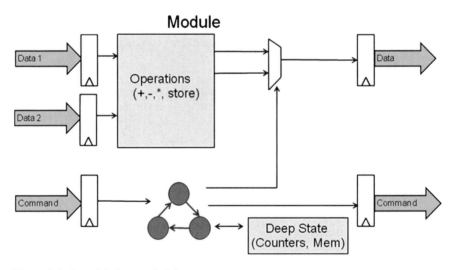

Figure 6-1 A module in canonical form.

Structure of the Canonical Design

The canonical design consists of:

- Data inputs (registered at the input of the module under test or at the output of the previous module).
- Control inputs (registered at the input of the module under test or at the output of the previous module). These should be coded as enumerated types; they issue commands or instructions to the control path of the design.

- Data outputs (registered at the output of the module under test)
- Control outputs (registered at the output of the module under test). These should be coded as enumerated types; they issue commands or instructions to the control path of the next module.
- A Data Path consisting of operations on data (arithmetic, logical, or storing in memory or a register).
- A Control Path consisting of:
  - a single state machine
  - deep state registers, counters, and memory

There is one sequential process for the state machine, and all assignments to control variables and data outputs occur in this process. All of the other sequential code (*always_ff* or equivalent statements) is limited to:

- Registers for inputs
- Internal registers for the data path

Note that the data path output registers are driven by the state machine. That is, the state machine determines when and what data path variables are written to the data path output registers.

Separating Data from Control

To put a design into canonical form, the first step in our analysis is to separate the data path from the control path. This separation is essential for verification, since we need to use different strategies for verifying each. To do this separation, we need to distinguish between control variables and data variables.

A control variable is a variable that appears in the condition in a conditional statement. For example:

```
if (write || read || (error == `DATA_ERROR)) begin
...
```

Here, the signals *write*, *read*, and *error* are control variables.

Control variables are control inputs, deep state variables, or functions of control inputs and deep state variables.

Data variables are variables that are not control variables.

From Figure 6-1, we see that data inputs can be combined. We can store data inputs in registers to create internal data variables. We can perform logical and arithmetic operations on data inputs and internal data variables. We can load data into output registers. As long as none of these results is used as a condition in a conditional statement, then all of these variables are elements in the data path, and are not part of the control path.

Verifying the Control Path: The State Machine

We start by focusing on verifying the control path for a design in the canonical form. This consists primarily of testing the state machine. We show that an extended code coverage measurement and give us a very good indication of how completely we have exercised the state machine.

The state machine code consists of lists of conditional actions:

```
case (state)
   STATE1: begin
      if (condition1) state              <= ...
      if (condition2) deep_state_n       <= ...
      if (condition3) control_output_n   <= ...
      if (condition4) data_output_n      <= ...
```

Where each condition can be reduced to the form:

A == B
A <= B
A >= B

A and B can be:

- A control input
- A deep state
- A constant or parameter
- A function of control inputs, deep state(s), and constants/parameters.

So in its most general form, a conditional state consists of one of the following:

if (f(control inputs, deep state)) deep_state_n <= g(control inputs, deep state);
if (f(control inputs, deep state)) control_output_n <= g(control inputs, deep state);
if (f(control inputs, deep state)) data_output_n <= g(data path);

Where *f* and *g* are arbitrary functions.
Note: the code for the state machine may consist of statements of the form

```
if (condition1) state <= foo;
else state <= bar;
```

but this is just equivalent to

```
if (condition1) state <= foo;
if (!condition1)state <= bar;
```

So our analysis applies to general conditional statements.

Given this structure, we can use (extended) code coverage to provide a large amount of information about the functional coverage of simulation – that is, how much of the function we have exercised.

Line Coverage

Achieving complete (100%) line coverage means we have entered every shallow state in the design.

Condition Coverage

Achieving complete (100%) condition coverage of statements of the form:
  if (condition) state <= ...
means we have executed every arc in the state machine, every possible way to get from one state to another.

Achieving complete (100%) line and condition coverage of statements of the form (including 100% coverage of the in-lined functions):
  if (f(control inputs, deep state)) deep_state_n <= g(control inputs, deep state);
means we have executed every arc into the accessible deep states.

Achieving complete (100%) condition coverage of statements of the form:
  if (f(control inputs, deep state)) control_output <= g(control inputs, deep state);
means we have executed every arc to the accessible output control states.

Achieving complete (100%) condition coverage of statements of the form:
  if (f(control inputs, deep state)) data_output <= g(data path);
means we have executed every way of updating the data outputs.

Conditional Range Coverage

Line and condition coverage are available in most code coverage tools. We now consider some additional coverage metrics that are not commonly available, but would be very useful for designs in our canonical form.

If we consider a conditional statement
  if (condition) ...

Where *condition* is one of:

A == B
A <= B
A >= B

Then it becomes important to know which values of A and B have been simulated to generate both the true and false conditions.

For instance, if the condition is "A < 5" then testing with the values of 4 and 5 provides more information than testing with 3 and 6 – it exercises the circuit at the exact point where the condition goes from true to false.

At the very least, then, we need a metric that indicates whether we have exercised the boundary condition of the conditional expression.

In addition, there might be cases where it is possible to create test cases that exercise all the possible values of A. It would be useful, then, to extend coverage tools to record which values of A were exercised.

Cycle Coverage

Line and conditional coverage indicates which (shallow) states and which arcs between shallow states have been covered. But this does not indicate which complete paths or cycles in the state machine have been covered. We need another coverage capability that measures what sequences of states have been exercised. In some designs this may be a very large space, and 100% coverage may not be possible (or even capable of being defined explicitly) but the metric combined with knowledge of the intended function of the design can provide valuable feedback on the quality of verification.

For instance, it would be useful to know that all paths through the state machine of length N have been exercised. It would be even more useful to be able to combine this with a formal proof that no (unique) paths of length greater than N are possible in the state machine.

Input State Coverage

The above code coverage metrics give us an excellent measure of how much of the control path (state machine) has been exercised. This in turn gives us some indication of which input states have been exercised as well. If all the possible paths through the state machine have been exercised, then we have most likely exercised all the interesting input states, and sequences of input states.

Nonetheless, it would be useful to have a separate measure of how much of the control input state space we have exercised. Such a metric could indicate what input states and what sequences of input states have been used during simulation.

Demonstrating that we have exercised all possible input states would give us a certain level of confidence in our verification suite. Demonstrating that we have covered all input state sequences of length less than some value M would give us even more information on the quality of the tests.

It would be even better if we could also formally prove that input state sequences of length greater than M do not produce any new behavior in the design.

Verifying the Data Path

There are two distinct problems in verifying the data path of a design.

1. Verifying that the state machine controls the data path correctly
2. Verifying that the data path implements the correct function or algorithm

The combination of our design strategy (using the canonical form) and extending code coverage help address the first problem. The next two sections address this aspect of verification.

Data Path Uniqueness

As we use code coverage to exercise all the possible ways to update the data outputs, we have a problem of observability. Consider the statement

```
if (A) data_output <= x + y;
else  data_output <= x * y;
```

Line coverage will tell us if we have executed both cases during simulation; but if x = y = 2, then in both cases the result written to *data_output* is 4. When A is true, we have a problem verifying that we actually wrote out x + y and not x * y. For this reason, it is useful to have the capability of determining if, when we update data outputs, the data output value uniquely identifies the data path calculation performed. It should be possible to extend code coverage tools to provide this analysis capability.

Data Range Coverage

As indicated earlier, it is not feasible to exercise an RTL data path completely if it has a large multiplier, or even a large adder. But such issues as overflow, rounding and saturation can be verified to some extent. To test these capabilities, we need a coverage metric that indicates whether boundary conditions for the data path – data values at or near the point where the data path is forced to round, saturate, or overflow – have been tested.

It would be useful to extend code coverage to include such a capability.

Verifying the Data Path Algorithm

The processes outlined in this chapter allow us to use (extended) code coverage to give a great deal of functional coverage – that is, a great deal of information about how completely the design has been exercised during simulation. However, there is a clear limitation on how much these methods can tell us about the data path itself.

For instance, if the data path implements a video algorithm, there may be no other way to verify the algorithm than running real video and looking at a display. The usual mechanism for this kind of verification is to develop a high level model (often in C) and use it to develop and verify the algorithm.

Once the algorithm is verified using a high level model and testbench, the key challenge is to show that the RTL version of the design is equivalent to the high level model – that it does, in fact, implement the algorithm. For complex algorithms, RTL simulation is not an effective way to do this. These algorithms can involve a number of large multipliers; as described earlier, this quickly leads to data paths that cannot be completely exercised in a finite amount of time.

The best solution to this problem is formal verification. Tools are becoming available that can prove the equivalence between a high-level, untimed model and an RTL implementation. As these tools mature, they should become a key tool in verifying the data path section of a design.

Summary

In typical RTL code, line and condition coverage give inadequate information on functional coverage. This is because

- Multiple concurrent sequential processes (effectively multiple concurrent state machines) produce a cross product space. The coverage of this cross product space is not measured by code coverage.
- Combinational code – multiple concurrent combinational processes – do not, in general, provide any indication of when their outputs are used functionally (that is, to change internal or output state). They are active in all states. There is no way to in-line them, to make their usage unique for each state. Thus, there is no way for code coverage to indicate how much of the functionality of the combinational process is being exercised.

Modules that have a single state machine that encompasses all the (non-data path) sequential code, and that use functions to encapsulate combinational code, and that have these functions called only by the state machine, are different. Under these conditions, code coverage indicates what functionality has been exercised, provided the coverage tool is extended to include:

- Functions in-lined in the state machine
- Conditional coverage indicating which values have been tested for both true and false results

- Coverage of input state space
- Cycle (path) coverage of the state machine
- Conditional range coverage
- Data uniqueness metrics
- Data range coverage

An open question for research is this: Can typical code be analyzed and automatically re-written so as to meet the requirements for code coverage to indicate functional coverage?

Chapter 7
Reducing Complexity in Data Path Dominated Designs

In Chapters 3 and 4, we discussed a control intensive design, the BCU. In this chapter, we discuss a data path intensive design: the discrete cosine transform (DCT) block in a JPEG design. We show how we recoded it to dramatically simplify the RTL.

Used in many video applications, the DCT is a two-dimensional transform performed on an 8 x 8 matrix of pixels. It is typically implemented as two one-dimensional transforms:

- A horizontal transform, which consists of multiplying the matrix, row by row, times a set of constants.
- A vertical transform, which consists of multiplying the modified matrix (the result of the horizontal transform), column by column, times a set of constants.

For the purposes of our discussion, we can ignore many of the details of the DCT. But it is useful to understand the basic structure and flow of the transform. To understand this structure, let's start with the C-language implementation of it.

As shown in Figure 7-1, we use three 8 x 8 arrays:

- The variable array *block*, which contains the 64 pixels being operated on
- The constant array *c* which contains the constant multipliers used in the algorithm
- The constant array *s* which indicates which of the partial products need to have their sign inverted

The algorithm itself consists of two steps: the first step processes the input array one row at a time (the horizontal transform), writing the result back into the array as it executes. The second step consists of processing this modified array one column up at a time (the vertical transform).

The C code for the DCT is shown in Appendix B. An abbreviated version is shown in Figure 7-1.

M. Keating, *The Simple Art of SoC Design: Closing the Gap between RTL and ESL*, DOI 10.1007/978-1-4419-8586-6_7, © Synopsys, Inc. 2011

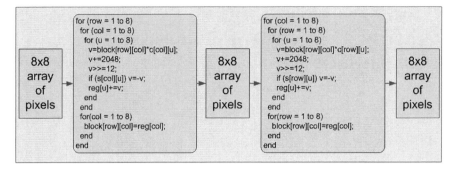

Figure 7-1 The DCT Algorithm.

In the hardware version, shown in Figure 7-2, the input array of pixels is replaced by a stream of pixels (one per clock cycle), and the output is also a stream of pixels (one per clock cycle). The results of the horizontal transform are stored in the DCTRAM, which provides inputs to the vertical transform. In the original RTL, the first (horizontal, or row-by-row) step is implemented as three separate modules:

- *mulh* which implements the multiplier – it creates eight products by multiplying the current pixel times each of the eight distinct values in the constant array *c*.
- *crossh* which routes each of the eight products to the appropriate element in the temporary array *reg_x*[7:0].
- *acch* which performs the inner accumulation loop on *reg_x*

The results are written into the DCT RAM.

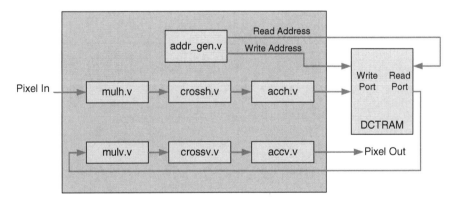

Figure 7-2 Block Diagram of the DCT.

The second (vertical, or column by column) step is implemented by the *mulv*, *crossv*, and *accv* blocks. It operates on data read from the DCT RAM.

The basic strategy for the hardware design relies on the characteristics of the constant array *c* that is used in the algorithm. In the C code, this array is defined as:

```
int c[8][8]={
{23168, 32144, 30272, 27248, 23168, 18208, 12544, 6400},
{23168, 27248, 12544, 6400, 23168, 32144, 30272, 18208},
{23168, 18208, 12544, 32144, 23168, 6400, 30272, 27248},
{23168, 6400, 30272, 18208, 23168, 27248, 12544, 32144},
{23168, 6400, 30272, 18208, 23168, 27248, 12544, 32144},
{23168, 18208, 12544, 32144, 23168, 6400, 30272, 27248},
{23168, 27248, 12544, 6400, 23168, 32144, 30272, 18208},
{23168, 32144, 30272, 27248, 23168, 18208, 12544, 6400}
};
```

Note that, even though there are 64 entries in the array, there are only eight distinct values. To perform the inner loop of the algorithm, the hardware implementation multiplies the target pixel with each of the eight distinct values. This is done in the *mulh* and *mulv* blocks. The *crossh* and *crossv* blocks then route the multiplication results to the appropriate accumulator, based on the particular (row, column) of the current pixel.

The accumulator blocks *acch* and *accv* perform the rest of the arithmetic (add by 2048, right shift, negation and summing). The *acch* block outputs one pixel at a time to the DCTRAM. When the DCTRAM has enough entries (one full column) then the *mulv* block starts reading from the DCTRAM and executing the column by column part of the algorithm.

The *addr_gen* module generates control signals for the entire DCT as well as addresses for the DCTRAM.

Problems and Limitations in the Original Code

The original design was coded in Verilog 95. As a result, it was limited to one dimensional arrays and unsigned data types only. It also contains a great deal of hand optimizations to guide synthesis.

The *mulh* code for the multiplication by eight constants shows this hand optimization. It starts by pre-calculating common constant multiplications as shift-and-add:

```
assign x3 = {x[10], x, 1'b 0} + {x[10], x[10], x};
assign x5 = {x[10], x, 2'b 00} + {x[10], x[10],
x[10], x};
assign x7 = {x, 3'b 000} - {x[10], x[10], x[10],
x};
assign x17 = {x[10], x, 4'b0000}+{x[10],x[10],x[10
],x[10],   x[10],x};
```

It then factors the constants into multiples of these pre-calculated values. Three of the eight calculations are shown below:

```
assign k7 = x17 + {x[10], x[10], x, 3'b 000};

assign k3 = {x7[13], x7[13], x7[13], x7[13],
x7[13], x7[13], x7[13], x7[13],x7} + {x5[13],
x5[13], x5[13],x5, 5'b 00000} + {x3, 9'b
000000000};

assign k0 = {x[10], x[10], x[10], x[10],
x[10], x[10], x[10], x[10],x} + {x5[13], x5[13], x5[13],
x5, 2'b 00} + {x5, 5'b 00000};
```

It turned out that all this effort to code the multipliers as shift and adds was completely wasted. When we re-coded this section to use simple multiplication (Example 7-1), we actually got better synthesis results (met timing, with smaller area). Modern synthesis tools really can find the optimal implementation for arithmetic operators.

We recoded the DCT, using SystemVerilog and focusing on three key points:

- minimizing the number of lines of code
- minimizing the complexity of the code – in particular, minimizing the amount of hand optimization in the code
- adding structure to the code to make it easier to understand

Minimizing Lines of Code

The original code for the DCT consisted of 1855 lines. In recoding the DCT, we were quickly able to reduce this to about 500 lines. We started by restructuring the code, combining the *mulh*, *crossh*, and *acch* modules into a single module called *dct_h*.

We use a similar strategy for the vertical blocks. Thus the structure of the code ended up as shown in Figure 7-3:

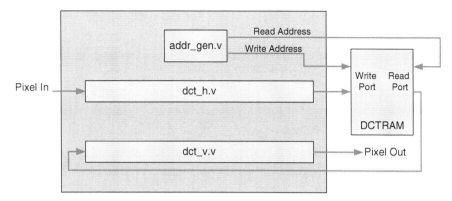

Figure 7-3 Block diagram of DCT code after first round of modification.

Converting the module declaration to the less verbose ANSI style and eliminating two module interfaces (and their verbose Verilog95 port declarations) resulted in *dct_h* being just a bit larger than *mulh*.

We then used signed data types, multidimensional arrays, and functions to simplify the code. The part of the code that replaced the *mulh.v* now looks like Example 7-1:

```
typedef bit [12:0] bit13;

function bit13 [7:0] get_y (bit signed [10:0] x);
    bit signed [15:0] k7;
    bit signed [16:0] k6;
    bit signed [20:0] k5;
    bit signed [21:0] k3;
    bit signed [19:0] k2;
    bit signed [21:0] k1;
    bit signed [18:0] k0;

    k7 = x * 25;
    k6 = x * 49;
    k5 = x * 569;
    k3 = x * 1703;
    k2 = x * 473;
    k1 = x * 2009;
    k0 = x * 181;
```

Example 7-1

```
   get_y[0]  = k0[18:6];
   get_y[1]  = k1[21:9];
   get_y[2]  = k2[19:7];
   get_y[3]  = k3[21:9];
   get_y[5]  = k5[20:8];
   get_y[6]  = {k6[16], k6[16:5]};
   get_y[7]  = {k7[15], k7[15], k7[15:5]};
   get_y[4]  = 0;
endfunction
```

Example 7-1   (continued)

Note that we used a function to add more structure to the code. We also eliminated the pre-computation of partial products – relying on synthesis to do the optimization.

In a similar way, we converted the *crossh* and *acch* into functions in the *dct_h*.

The structure of the *dct_h* is shown graphically in Figure 7-4. The code is outlined in Figure 7-5. It consists of two sequential processes and three functions. The bodies of the processes and functions have been replaced by comments to make the overall structure easier to see.

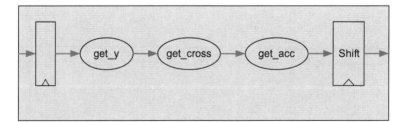

Figure 7-4   Diagram of code showing the functions used to encapsulate combinational code.

```
always_ff @(posedge) begin
  register input pixel

function get_y ;
  does initial math on registered input

function get_crossh;
  calls get_y() and maps result to right location
  in matrix

function get_acc_value;
  calls get_crossh and calculates the accumulated
  value

always_ff @(posedge clk)
  calls get_acc_value(), registers the result and
  shifts out
```

Figure 7-5 Outline of modified code.

The input pixel is clocked into a flip-flop, then three functions (*get_y, get_cross, get_acc*) perform the inner loop of the calculation, and the result (eight pixel values) are clocked into a shift register (after one row or column has been processed). The shift register is then clocked out one pixel at a time.

We make the following observations about this code:

1. The module declaration is much simpler than the original. Much of this is due to using the ANSI style port declaration. But in addition, we restructured some of the communication between *addr_gen* and *dct_h* to reduce the number of control signals.

2. We now use functions instead of combinational code. This allows us to declare the intermediate values as local variables within the function. This simplifies the code and makes review and analysis easier.

3. By using signed types, we were able to reduce the amount of manual sign extension.

4. By using multidimensional arrays, including defining (the return value of) the functions as multidimensional arrays, we were able to simplify the code.

5. There is no state machine in the design. The input is registered and the output is registered, but no control state is saved.

The code for the final sequential process in *dct_h* looks like Example 7-2 (with reset code removed for easier reading):

```
always_ff @(posedge clk) begin
  if (en) begin
     reg_x <= get_acc(sgn, sel, load, reg_x);
     if (load) sh <= reg_x[7:1];
     else for (int i=1;i<7;i++) sh[i]<= sh[i+1];
  end
end

always_comb dct_h_out = load ? reg_x[0] : sh[1];
```

Example 7-2

So the accumulator registers (*reg_x*) call the function *get_acc*, which calls the function *get_cross*, which calls the function *get_y*. The only combinational logic in this version of the code is the assignment of *dct_h_out*.

Note that the final sequential process is followed by a combinational assignment. This is one of those cases where a combinational output has to be driven directly out of the module to meet overall system timing.

Table 7-1

	Original Verilog95 Version	First SystemVerilog Version
Lines of code	1855	519
Files	8	4
Sequential processes	5	5
Combinational processes	5	6
Assign statements	144	6
Functions	0	8
Tasks	0	0

Table 7-1 summarizes the progress we made in reducing the complexity of the DCT code. In addition to the reduction in the lines of code, the biggest difference is in the number of assign statements. We have replaced almost all the assign statements with functions, replacing unstructured code with structured code.

Other Versions of the Code

After making the modification described above, we created some other versions of the DCT code.

Task Version

In the task version, we decided to add more structure to dct_h and dct_v. Earlier in this book, we observed the difficulty in encapsulating sequential code, in a way equivalent to the *function*. In a control-dominated design, we can use a state machine to structure sequential code. In the case of the DCT, there is no state machine. Instead, we used *tasks* to structure the sequential code.

In the task version of the code, we used a task to register the inputs and outputs, and do the output shifting. We then used a single sequential process to call these two tasks. The code for the task version of dct_h.v is outlined in Figure 7-6.

```
task update_input();
  register input pixel

function get_y;
  does initial math on registered input

function get_crossh;
  calls get_y() and maps result to right location
  in matrix

function get_acc_value
  calls get_crossh and calculates the accumulated
  value

task update_sr();
  calls get_acc_value(), registers the result and
  shifts out

//------------------- MAIN -----------------------
always @(posedge clk) begin
  update_input ();
  update_sr ();
end
```

Figure 7-6 Outline of the task version of the code.

Some notes about the tasks:

1. The tasks use non-blocking assignments. The result is that the new code is exactly equivalent to the old code. It is just more structured.
2. The added structure makes analysis of the module easier. The sequential process acts like the *main* of a C program. We start by looking at what this process does; everything else in the code – the tasks and functions – exists only to support the *main* sequential process.

3. The registers controlled by the two tasks (the variable x in *update_input*, and the variables *reg_x* and *sh* in *update_sr*) are global – declared at the top level of the module – rather than local.

Table 7-2

	Original Verilog95	First System-Verilog	Task Version
Lines of code	1855	519	544
Files	8	4	4
Sequential processes	5	5	3
Combinational processes	5	6	3
Assign statements	144	6	1
Functions	0	8	10
Tasks	0	0	6

Table 7-2 summarizes the progress made so far. Note that in addition to using tasks to add structure to the sequential code, we have also migrated more of the combinational code into functions. We have now started refactoring the *addr_gen* block in addition to *dct_h* and *dct_v*. Most of the improvement in the combinational code is a result of the work on the *addr_gen* block.

Note also that there was a slight increase in lines of code for the task version. This growth is due to several factors:

- the overhead of a *task/function* compared to an *assign* statement (*task/endtask* and *function/endfunction* in a total of 8 places)
- a few more comments in the task version
- a few more lines of white space to improve readability of the code

But we more than made up for this 5% growth in the next round of code reduction.

More Code Size Reduction

In looking at the code for the task version of the DCT, we realized that we had reduced the code size, and improved the structure of the code, to the point where we could start combining four modules into one module.

The main reason to use multiple modules was to keep each file down to a reasonable size – five pages or less – so we could read and analyze each module without getting overwhelmed. We found that having multiple modules – and the overhead of multiple port lists – was no longer an advantage. So we combined the modules into a single module, and were able to reduce the total DCT code to about 280 lines. This is less than 5 pages, which we regard as the maximum recommended length for a module.

By combining the four modules into one module, we could use a single sequential process to invoke all the tasks. This single sequential process now acts as the equivalent of *main* in C programs. To review and analyze the DCT code, we just start at this process and follow the task calls. This is now truly structured RTL.

The full code for the 1-file version of the DCT is included in Appendix B.
Figure 7-7 gives an outline of the code.

```verilog
//------ from addr_gen.v ------------

task update_addr();
  manages the address to RAM that stores
  intermediate results, calling
  increment_addr() twice;

function automatic increment_addr;
  write (horizontal) and read (vertical)
  addresses to RAM; this is
  called twice, so we declare it an
  automatic function


function automatic get_sgn ();
  calculate the sign of the constant used in
  get_y and get_y_vert

//-------------- horizontal from dct_h.v --------
function get_y;
function get_crossh;
function get_reg_x;
task update_sr;


//-------------- vertical from dct_v.v------------
function get_y_vert ;
function get_crossv ;
function get_reg_xv ;
task update_srv ;


//-------------------- MAIN ------------------------
task main ();
  horiz_in <= pixel_in;
  update_addr();
  update_sr();
  vertical_in <= ram_out;
  update_srv;
endtask

always_ff @(posedge clk) main();
```

Figure 7-7 Outline of the 1-file version of the DCT.

Table 7-3 shows the progress made in the various versions of the DCT code.

Table 7-3

	Original Verilog95	First SystemVerilog	Task Version	1-File Version
Lines of code	1855	519	544	278
Files	8	4	4	1
Sequential processes	5	5	3	1
Combinational processes	5	6	3	2
Assign statements	144	6	1	0
Functions	0	8	10	8
Tasks	0	0	6	4

The final count of lines of code is quite remarkable. The reference C model – at the highest level of abstraction, and using integer arithmetic – is about 125 lines of code. Thus, with some effort – and taking advantage of the language features now available – we were able to reduce the code size of the design by a factor of about 5x, to where the RTL was only a little over 2x the size of the reference C code.

Untimed Version

One advantage of the 1-file version of the DCT is that the clock only appears once – in the single sequential process at the end of the module. We can then make some minor changes, and use the code to create an untimed model of the DCT. In the untimed version, we had to make the following changes:

1. The testbench had to be modified. Instead of simply instantiating the DCT and generating a clock, it explicitly calls the main task in the DCT for each pixel being processed.
2. The lone remaining sequential process was deleted.
3. The non-blocking assignments in the tasks have been changed to blocking. This is necessary to assure the correct sequence of execution. (That is, the discipline established by clocking is replaced by the discipline of the order of assignments). In two cases, we had to re-order some statements to support this discipline.
4. In the timed code, it takes two cycles to read and write the DCTRAM. In the untimed model, this is done in zero time. As a result, the exact timing of some control signals changes. The variable *index* is used in the code to coordinate various activities. In several cases we had to change the way index was used, such as changing a condition *if (index == 0)* to *if (index == 7)*.

Figure 7-8 gives the outline for the code for the task version. Figure 7-9 shows how the testbench invokes the dct.

The objective of this exercise was to see if the same code could be used for both an untimed and a timed model of the design.

The advantage of an untimed model is that it simulates faster and is easier to understand and debug. The execution of the untimed model involves no concurrency. Like a C model, it executes in a linear fashion, completing one calculation before starting another.

The fact that we had to make changes other than simply eliminating the clock was disappointing. Nonetheless, we have shown that RTL can be coded in a structured format that makes for simple, if not yet automated, conversion between untimed and timed versions.

```
//------ from addr_gen.v ------------
task update_addr();
function automatic increment_addr;
function automatic get_sgn ();

//-------------- horizontal from dct_h.v ------
function get_y;
function get_crossh;
function get_reg_x;
task update_sr;

//-------------- vertical from dct_v.v-------
function get_y_vert ;
function get_crossv ;
function get_reg_xv ;
task update_srv ;

//------------------- MAIN -------------------
task main ();
 horiz_in = pixel_in;
 update_addr();
 update_sr();
 vertical_in = ram_out;
 update_srv;
endtask
```

Figure 7-8 Outline of the untimed version of the DCT.

```
initial begin
  for (int i=0;i<MAX_TEST;i++) begin
      get_input_pixel();
      u1.main();
  end
  $finish;
end
```

Figure 7-9 (Partial) test bench for the untimed version of the DCT.

Experimental Versions

Based on previous versions of the code, we developed two experimental versions of the DCT. They both used integer arithmetic – all variables are declared to be type *int* (as in the C version).

From the 1-file version, we created the ExpClk version, which replaced the bit-level arithmetic with integer arithmetic, and used a less efficient (but simpler) way of handling the constant array.

Starting from the untimed version, we developed the Exp version, which again used integer math and the simpler constant handling.

Figure 7-10 shows a fragment of the code. This task is used by both the Exp and ExpClk versions, either called explicitly by the test bench (Exp verion) or executed at each clock by an always_ff process (ExpClk version).

Some notes on the code:

1. All the variables are declared as type integer. This results in faster simulation, but of course synthesis produces a design with much larger area.
2. The constant matrices *s* and *c* are used directly as in the C code. We again rely on synthesis to do efficient mappings, rather than the manual mapping done in the function *get_crossh*.

```
int reg_x [7:0];
int x,v;
int sh [7:0];

const int c[8][8]='{
'{23168,32144,30272,27248,23168,18208,12544,6400},
'{23168,27248,12544,6400,23168,32144,30272,18208},
'{23168,18208,12544,32144,23168,6400,30272,27248},
'{23168,6400,30272,18208,23168,27248,12544,32144},
```

Figure 7-10 Integer version of the code.

```
'{23168,6400,30272,18208,23168,27248,12544,32144},
'{23168,18208,12544,32144,23168,6400,30272,27248},
'{23168,27248,12544,6400,23168,32144,30272,18208},
'{23168,32144,30272,27248,23168,18208,12544,6400}
};
int s[8][8]='{
'{0, 0, 0, 0, 0, 0, 0, 0},
'{0, 0, 0, 1, 1, 1, 1, 1},
'{0, 0, 1, 1, 1, 0, 0, 0},
'{0, 0, 1, 1, 0, 0, 1, 1},
'{0, 1, 1, 0, 0, 1, 1, 0},
'{0, 1, 1, 0, 1, 1, 0, 1},
'{0, 1, 0, 0, 1, 0, 1, 0},
'{0, 1, 0, 1, 0, 1, 0, 1}
};
task update_sr();
 if (index == 0) reg_x = '{0,0,0,0,0,0,0,0};
 else x = pixin - 128;
 for(int u=0;u<8;u++)begin
  v= x *c[index][u];
  v+=2048;
  v = v >>>12;
  if(s[index][u]) v=-v;
  reg_x[u]+=v;
 end
 if (index == 7)
  for (int i=0;i<8;i++) sh[i]= reg_x[i];
 else for (int i=0;i<7;i++) sh[i]= sh[i+1];
endtask
```

Figure 7-10 (continued)

Simulation Results for the Different Versions of the DCT

Reference Versions

We used two reference versions of the code in running simulation.

1. The original C code. Coded for speed and simplicity at a high level. The code is shown in Appendix B.
2. A Reference SystemVerilog version. For this version, we copied the C code into SystemVerilog and made only those changes necessary to get it to compile as a SystemVerilog design.

To convert the C code to legal (but not synthesizable) SystemVerilog code, we had to make the following changes:

- change the { } to begin/end.
- Add apostrophes at the appropriate points in the declaration of the constant arrays *c* and *s*.
- Change *void dct* to *function dct*.
- Change *reg* to *reg_x*, since *reg* is a reserved work in SystemVerilog.

With these rather trivial changes, the code now compiled and ran in the VCS simulator. Figure 10-2 (in Chapter 10– Raising Abstraction Above RTL) shows the converted code.

We then ran simulations for the different versions of the DCT. In each case we ran the same pixel streams, with the simulation run consisting of five million pixels. The results are presented in Table 7-4:

Table 7-4

Name	Description	Runtime
Verilog 95	Original Verilog95 version of dct	115 sec
Function	Original converted to SystemVerilog using functions	119 sec
Task	Function version plus sequential processes as tasks	118 sec
Untimed	Task version modified to run untimed	77 sec
ExpClk	Experimental version using integer math	19 sec
Exp	Experimental version using Untimed structure, integer math	18 sec
Ref SV	C code translated directly into SystemVerilog	6.9 sec
Ref C	Original C reference version of dct	2.5 sec

Conclusions: The C was clearly the fastest version. The Int version, virtually identical to the C version, ran somewhat slower because SystemVerilog simulation performs run-time bounds checking, which C does not.

The Verilog95, (SystemVerilog) Function, and (SystemVerilog) Task versions all ran at essentially the same speed.

The Untimed version ran considerably faster than the three timed versions, but slower than the Reference C and Reference SystemVerilog versions. Clearly eliminating the clock provides some speed improvement, but using bit rather than integer arithmetic resulted in slower simulation than the reference (integer) versions.

The Exp (integer) version shows a significant speed-up over the bit-based versions. This implies that the biggest speed-up in SystemVerilog simulation comes with integer arithmetic. This makes sense because the simulator can use native integer types for integer simulation; bit-based simulation requires more complex and slower data representation.

The Verilog95, function, and task versions are all synthesizable. The ExpClk version is also synthesizable, but because it uses integer math, the area is much larger than for the other synthesizable versions. The area results are given below. All versions met timing.

Synthesis Results

We synthesized the three versions of the DCT using a commercial 65nm library and a 100MHz clock rates. Table 7-5 summarizes the synthesis results.

Table 7-5

Name	Description	Area(gates)
Verilog 95	Original Verilog95 version of dct	16,308
1 File	Original converted to SystemVerilog using functions, tasks, and in a single file with one sequential process	15,400
ExpClk	Experimental version using integer math	43,874

We then re-ran synthesis for the Verilog95 and 1-File versions at a 250MHz clock rate. The results are shown in Table 7-6.

Table 7-6

Name	Description	Area(gates)
Verilog 95	Original Verilog95 version of dct	25,818
1 File	Original converted to SystemVerilog using functions, tasks, and in a single file with one sequential process	18,000

The fact that the 1-File version is smaller than the original Verilog95 version is mostly due to the fact that the multipliers in the original version were coded as shift-and-add. This form of code constrained the synthesis tool and produced a sub-optimal result. In the 1-File version we left the multipliers as multipliers, allowing the tool to explore a broader range of implementations.

Gates per Line of Code

There has long been a rough estimate that Verilog code produces about 5 gates per line of code (for control dominated designs) to about 10 gates per line of code (for data path dominated designs). The Verilog95 version of the DCT meets this estimate, at roughly 9 gates per line of code.

Many observers have stated that to get an order of magnitude improvement over these (gates per line of code) numbers requires C-based high level synthesis. But in fact the final RTL version of the DCT achieved about 58 gates per line of code, an improvement of almost 6x.

We also took the original C code and used a C-based high level synthesis tool to generate RTL (and then synthesized to gates). To achieve the QOR (specifically the area) of the RTL versions, we had to add a significant amount of code to the original C version (mostly details for handling the constant multiply efficiently). The final (synthesizable) C version was about 263 lines of code, for about 61 gates per line of code.

Thus, our final RTL reached about the same code density (gates per line of code) as the C-based version. Although there are other advantages of synthesizing from untimed C models, these results suggest the following: The big leap forward from where we are today to a 10x improvement is not necessarily a leap from Verilog to C. Simply moving from today's code to well-written SystemVerilog code can produce a dramatic improvement.

Formal Verification

To verify that the different versions of the dct were functionally equivalent, we initially used a simple test bench and simulation.

Later, we tried a more sophisticated verification strategy using formal verification. We could not use Formality, because Formality cannot check the equivalence of an untimed model to a timed model.

Instead, we used Hector, an experimental formal verification tool that can compare (untimed) C models to RTL. Using Hector, we were able to prove that the Task implementation of the DCT was equivalent to the (untimed) C model.

Canonical Design

When we look at the Canonical Design for a control-dominated design, it looks like Figure 7-11.

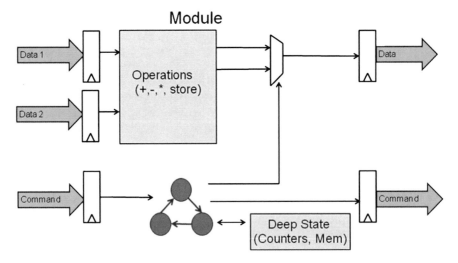

Figure 7-11 Canonical form for a digital design.

But for our data-path intensive design, the model is slightly different. In the DCT design, the model looks more like Figure 7-12:

Module

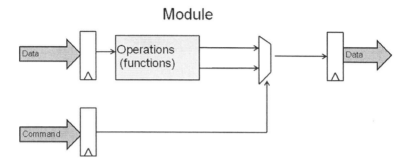

Figure 7-12 DCT has no state machine, so it is a simple subset of the canonical design.

Without the extensive control logic, we do not have a state machine or a concern about deep and shallow state. Instead, commands come into the module and directly control the data path. It is the data path itself that requires the most attention: partitioning it into functions to make the analysis simpler.

Summary

As in the case of a control-dominated design, we have found that a structured coding and design style reduces the complexity and number of lines of code in a design. In particular, we can reduce the number of sequential processes to a minimum.

Using functions and tasks to add structure to the code produces a canonical design for a data path module, and facilitates analysis and verification. As in the case of the example in Chapter 3, we were able to have a single sequential process per module. As a result, each module has the equivalent of a *main* – a single point of entry, where a systematic analysis of the module can begin.

We have also shown that using these techniques – and the features available in the synthesizable subset of SystemVerilog – we could reduce the code size dramatically, without adversely affecting simulation time or synthesis results.

Chapter 8
Simplifying Interfaces

In previous chapters, we described how to simplify modules. In this chapter we describe how to simplify the interfaces between modules. In doing so, we use the same basic approach used in addressing module design:

- Reducing the number of lines of code
- Reducing the complexity of the design by minimizing the state space
- Add structure to the code to make analysis of the code easier

We start with a trivial example which, while small, nonetheless shows one principle we will use.

Command-based Interface

Consider the following code:

```
module dct_h (
   input bit read,
   input bit write,
         .
         .
         .
endmodule
```

M. Keating, *The Simple Art of SoC Design: Closing the Gap between RTL and ESL*, DOI 10.1007/978-1-4419-8586-6_8, © Synopsys, Inc. 2011

Read and write are both control inputs (not data). The input state space for this module is $2^2 = 4$ states. But we may not support doing both read and write at the same time. In this case, the real state space is three, as expressed by the following code:

```
typedef enum bit {NOP, READ, WRITE} rw_type;

module dct_h (
  input rw_type rw,
      .
      .
      .
endmodule
```

Now we have made it clear that either read or write or nothing can happen – but read and write can never be active at the same time. We are moving from an interface consisting of wires to an interface consisting of commands.

In this case we have reduced the input state space of the design from four to three – not a huge win numerically. But when applied to a real design, the reduction can be dramatic.

The general model is shown in Figure 8-1:

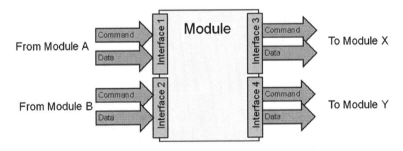

Figure 8-1 Module interfaces using command and data ports.

The module has one (input) interface from each module that drives it and one (output) interface to each module that it drives.

To minimize the input/output state space, we explicitly separate data from control, and then declare the control inputs as an enumerated type. The complexity (for a single interface) then goes from 2^n, where n is the number of wires, to N, where N is the number of commands listed in the enumerated type.

Although the command interface is drawn (and thought of) as unidirectional, it often is a handshake interface. There is a preferred handshake for interfaces between modules: the FIFO.

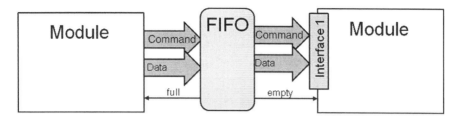

Figure 8-2 A FIFO is the preferred interface between modules.

The FIFO provides a well-defined handshake interface; it also provides timing isolation between the two modules. The sending module does not need to know if the receiving module is ready for data; it just monitors the FIFO status, and if it is not full, it can deliver a new command and data. Similarly, the receiving module does not have to know anything about the internal timing of the sending module; it just monitors the empty signal from the FIFO, and knows when new command/data are available.

In those cases where a single register is used between modules, there are two cases:

1. The receiving module is guaranteed always to be ready to receive data/command
2. The receiving module is not always ready to receive data/command, and needs to throttle back the sending module

In the first case, a simple register interface suffices: this is just equivalent to a 1-deep FIFO with full always negated.

In the second case, it is best to use a 1-deep FIFO. There are many ways to create an ad-hoc handshake across modules, but the FIFO is a standard handshake that all engineers understand.

Example: CPU Pipeline

The classic example of a command-based interface is the CPU pipeline. A simple 5-stage RISC processor[12] might have an Instruction Decode (ID) to Execute Stage (EX) interface like Figure 8-3.

The ALU receives a command (IRp1, or Instruction Register piped one stage) and data from the ID stage. The Data consists of:

• Program Counter (for jumps, which require calculation of the target address): register PC_p1 in the diagram.
• Outputs from the Register File (for normal operations, like shift or add): registers A and B in the diagram.

- Immediate (for instructions that encode one argument in the instruction itself): register IMM in the diagram.
- The ALU output from the previous cycle (for forwarding, to avoid unnecessary pipeline stalls): register RSLT in the diagram.

The opcode (IRp1 in the diagram) is the command that tells the ALU what to do with this data.

A simplified diagram might look like Figure 8-4.

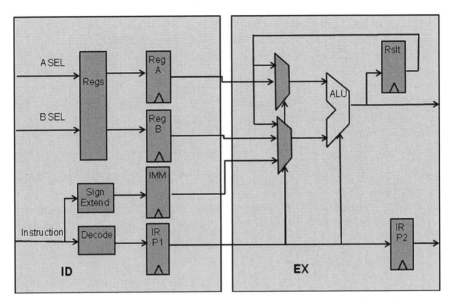

Figure 8-3 Instruction Decode and Execute stages of a pipelined processor.

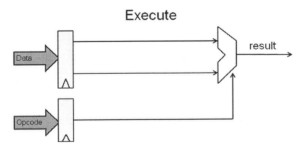

Figure 8-4 Simplified diagram of the EX stage.

The concept of an opcode – an enumerated list of commands that tell the module how to manipulate the data – is the concept we are generalizing to all module interfaces.

In an actual CPU design, there are more inputs than just the opcode.

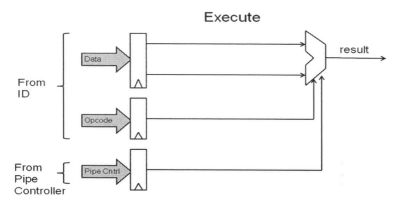

Figure 8-5 EX stage with control inputs from Pipe Controller.

As shown in Figure 8-5, additional signals – stall and forward, for example – typically come from other stages or from a central pipeline controller. In one design we have seen, there were eighty-nine opcodes and nine independent control signals, for an input control state space of

$$89 * 2^9 = 45,568 \text{ or less than } 2^{16}.$$

This is a large state space, but quite manageable in automation terms, if not in human terms. To simulate all possible input states, executing at 2,000 instructions per second (a reasonable simulation speed for a simple processor) would take less than a minute.

Example: BCU

The BCU described in Chapter 3 is a bit more complicated, because it has interfaces to multiple modules. Its input declaration (control inputs only – not data) looks like Example 8-1.

```
input bit          bium_dma_done,
input bit          bium_error,
input bit          bium_wdata_pop,
input bit          bium_rdata_push,
input bit          csr_disable_bium_abort,
input bit          bcu_fifo_busy,
input bit [1:0]    bcu_txq_empty,
input bit          bcu_rxq_empty,
input bit          bcu_dwq_empty,
input bit          bcu_dfq_empty,
input bit          bcu_rxf_empty,
input bit          bcu_dwf_empty,
input bit          dcub_txfnum_vld,
input bit          ccub_rxf_stop,
input bit [1:0]    ccub_txf_stop,
input bit          ccub_rxf_resume,
input bit          ccub_txf_resume,
input bit          ccub_resp_rdy,
```

Example 8-1

This is 20 signals, for a state space of $2^{20} = 1,048,576$. But it can be made smaller, since many states are mutually exclusive, and can be combined into an enumerated structure.

For example, the first three signals (control signals from the *bium*) are not independent. Only one can be active at a time. Thus, they can be combined into a single enumerated type with four values:

```
typedef enum bit [1:0] {bium_nop, bium_dma_done,
bium_error, bium_wdata_pop } bium_cmd ;
```

Similarly, the five inputs from the ccub:

```
ccub_rxf_stop, ccub_txf_stop, ccub_rxf_resume,
ccub_txf_resume, ccub_resp_rdy,
```

can be combined into:

```
typedef enum bit { ccub_nop, ccub_rxf_stop, ccub_
txf_stop_0, ccub_txf_stop1, ccub_rxf_resume, ccub_
txf_resume, ccub_resp_rdy, } ccub_cmd ;
```

Just making these two simple changes reduces the input state space to:

$$4 * 2^9 * 7 = 14,336$$

We have then reduced the input state space from over 1,000,000 to about 14,000, an improvement of 73x, or about 2 orders of magnitude. More importantly, the input state space is now structured: there are four bium commands and seven ccub commands.

Example: USB

CPU design is a mature science. Passing the opcode (rather than random wires) from stage to stage has been well established for many years. In the BCU example, we are dealing with a recent design that incorporates modern concepts of interface design. That is, the designer purposely minimized the complexity of the interface to the degree possible in Verilog 2K. The improvements we were able to make stem directly from our ability to use SystemVerilog.

For an example of how the interface state space can get out of control, we now look at an older design, a module from a USB interface designed a decade ago.

Consider the following set of (control) input declarations in Example 8-2.

There are 53 wires in the control inputs, for an input state space of:

$$2^{53} = 9,007,199,254,740,992$$

In human terms this is essentially infinite. But many of these input control signals are mutually exclusive, and could be combined into enumerated states, dramatically reducing the input state space. This kind of analysis could be done easily by the original designer. But for anyone else, this analysis would be a challenging task indeed.

This case is an excellent example of how design intent becomes lost when the input state space is not carefully designed. The original designer of this code certainly understood which signals were mutually exclusive and understood the intended input state space, which was probably no larger than that of the BCU -- on the order of 2^{20}. But by coding the inputs as he did, the original designer completely hid which were the intended input states and which input states were never intended to occur.

```
//Control inputs
input   rst_phyclk;
input   phy_clk;
input   se_hs_hnsk_on;
input   sr_bus_idle;
input   se_send_chirp;
input   phy_rxvalid;
input   phy_rxvalidh;
input   phy_rxerr;
input   phy_rxactive;
input   phy_txready;
input   epi_cur_dtsync;
input   epi_dec_desccmd;
input   app_setdesc_sup;
input   app_synccmd_sup;
input   ubl_cntrlep;
input   ubl_bulkep;
input   ubl_cmdstate_intrd;
input   ubl_cmdstate_sts;
input   ubl_ep_ok;
input   ubl_ext_cyc;
input   ubl_ep0_cyc;
input   ubl_isoep;
input   ubl_send_stall;
input   ubl_send_nak;
input   ubl_rx_err;
input   ubl_send_nyet;
input   ubl_cntrl_hshk;
input   ubl_get_status;
input   ubl_int_rd;
input   se_usbreset;
input   sync_wrcmd_sts_ok;
input   sync_srcbuf_empty;
input   app_phyif_8bit;
input   ubl_se0_nak;
input   ubl_test_j;
input   ubl_test_k;
input   ubl_test_pkt;
input   sr_remotewakeup;
input   sync_dev_discon;
input   line_state_se0;
```

Example 8-2

```
input   app_nz_len_pkt_stall;
input   app_nz_len_pkt_stall_all;
input   app_enable_erratic_err;
input   [2:0] hs_timeout_calib;
input   [2:0] fs_timeout_calib;
input   [1:0] se_enum_speed;
input   app_scale_down;
input   byteif_txvalid;
```

Example 8-2   (continued)

Impact on Verification

Minimizing the input state space of the design has a significant impact on making the design easier to understand. But it also has a significant impact on verification.

For designs such as the USB module described above, with an input state space of 2^{53}, verifying the entire set of possible inputs to the design is not practical. At a simulation speed of 10,000 cycles per second (a very fast simulation environment), it would take more than 28,000 years just to stimulate the module with every possible input state. Or, if we used 28,000 cpu's, we could do it in one year. Thus, completely verifying just the input state space of the design is impractical.

On the other hand, for designs with an input state space on the order of 50,000, such as the execute stage in a CPU, the numbers are quite different. Even at 2,000 cycles per second, we can drive all possible input states in less than a minute. Clearly, a much larger portion of the behavior of the module can be verified in an acceptable amount of time.

The key point here is that without carefully managing the input state space, the very concept of complete verification is impossible.

Separating Data and Control

To minimize and structure the input state space, it is necessary to separate data and control. The advantage of doing this has already been described in the chapter on verification.

In some cases, a single command interface may consist of both data and control. For instance, a packet read from a FIFO may contain both data and information as to how the data should be handled. In this case we may want to use a struct to clearly delineate the separation between command and data. Consider, for example, the following declaration in the BCU:

```
struct packed{
  bit          tr ;    //      Control:          Transmit
  direction
  bit          ds ;    // Control:  Descriptor
  bit          rr ;    // Control:  Response required
  bit [1:0]    tag ;   // Data: Tag ID
  bit [31:0]   addr ;  // Data: DMA address
} dmareq;
```

In this case, we use comments to make explicit which fields in the structure or control and which fields are data. In the future, it might be useful to have the capability within SystemVerilog to tag fields as either control or data. Making the distinction between control and data inputs an explicit capability of the language would facilitate manual and automatic analysis of the source code.

General Connectivity

So far, we have talked about the internal state space (of a module) and the input state space (of a module) as metrics of complexity. Another important measure of complexity is the connectivity between modules.

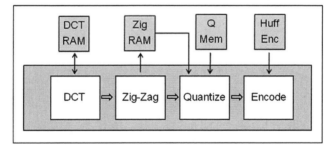

Figure 8-6 Simplified diagram of a JPEG core.

In the case of a data path intensive design such as a JPEG core, shown in Figure 8-6, the architecture is defined in terms of data-processing functions such as dct or quantizer. The modules are arranged to fit this architecture, with each module having an interface to the preceding data block and to the succeeding data block.

The result is that each module communicates closely with two other blocks, but has little direct communication with other blocks. This results in a fairly simple connectivity graph. More importantly, it means that changes to any one block are unlikely to ripple through and affect all the other blocks in the design – rather, such changes are likely to impact, at most, the blocks two neighbors.

In a control-dominated design, such as the USB core shown in Figure 8-7, the architecture is shown in a series of layers, mirroring the layers of the communication protocol.

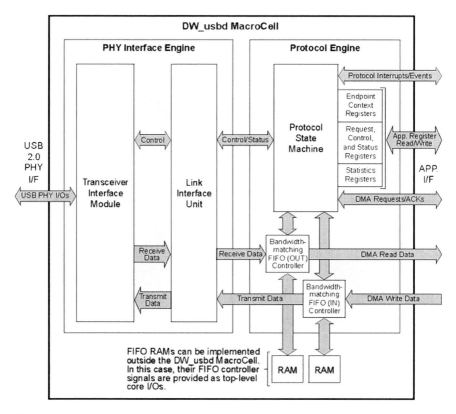

Figure 8-7 Block diagram of the USB.

In this case, the individual modules again tend to communicated to their nearest neighbors, with no direct connection, for instance, between the Transceiver Interface Module and the Protocol State Machine.

Both of these designs have architectures that limit interdependencies between modules. A poorly architected design would look something like Figure 8-8.

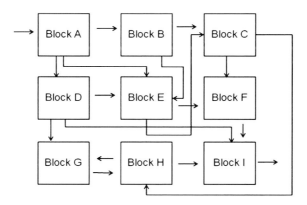

Figure 8-8 A poorly architected design with excessive connectivity.

Here each module communicates directly with (almost) every other module. The connectivity graph is almost complete instead of sparse. The result is that changing any one module (fixing a bug, or adding a feature) is likely to require modifying many, if not all, the other modules. This can lead to a design that is virtually impossible to maintain.

Concurrency and Analysis

In this chapter, as in others, we have tried to show techniques for minimizing concurrency in order to facilitate analysis of a design. By using a single state machine within a module, we minimize concurrency within the module.

By using a FIFO between modules, we minimize the impact of concurrency between modules. Although the two modules operate concurrently, the FIFO isolates much of this concurrency. Each module can be analyzed independently, and be thought of as receiving and delivering a data stream to its neighbors. Even though the modules operate concurrently, there is no need to analyze them concurrently, except to make sure the FIFOs are of the appropriate depth to meet performance objectives.

This appropriate FIFO depth is often difficult to determine analytically, but can be determined by simulation. For small designs, we can run RTL simulation with realistic workloads to determine the required FIFO size. For larger designs, we may have to use high level modeling: transaction level models that are fast enough to allow us to run real workloads and that are accurate enough to reflect delays and latencies in the system.

Total State Space

If we look again at the BCU design, we see that by re-structuring the design we reduced the internal state space from about 2^{56} to about 2^4. This was described in Chapter 3.

In this chapter, we reduced the input state space of the BCU from about 2^{20} to about 2^{14}. The total state space of the design is the cross product of these two state spaces, so we have reduced the total state space of the BCU from:

$$2^{56} * 2^{20} \qquad \text{to} \qquad 2^4 * 2^{14}$$

or from essentially infinite to about 262,000. Although this state space is still large, it is now small enough that we can consider verifying a large part (if not all) of it.

Summary

In previous chapters, we showed that careful, structured design of control-dominated and data-path dominated code can result in much simpler designs. In this chapter, we showed that the same principles apply to interface design.

Interfaces, such as a FIFO interface, that decouple the two modules can greatly reduce the complexity of the overall design. In a tightly coupled design, the complexity of the design is the cross product of the complexity of the individual modules. In a decoupled design, the complexity is the complexity of the interface. With a well-structured interface, this can be orders of magnitude smaller.

A well-structured interface uses *structs* to group wires into fields and to separate these fields into data and control words. So we move from thinking about wires and bits to thinking about control, data, and transactions. Enumerated types reduce the state space complexity of the control word in the interface.

Chapter 9
Complexity at the Chip Level

This chapter extends some of the concepts introduced in earlier chapters, and applies them to the IP, subsystem, and chip level.

A central point of this book so far is to develop useful metrics of complexity in design, and then use these metrics to identify:

- Methods for simplifying designs so they are more likely to be functionally correct
- Opportunities for tools – both synthesis and verification – to help develop and verify designs better
- Language constructs that would make it easier to develop simpler designs

All of the techniques described so far are responses to what the metrics tell us.

The key metrics proposed include:

- Lines of code
- Internal (shallow) state space
- Input state space (we ignore output state space since it is someone else's input state space, but this may be a mistake)
- Structure – how many objects in the design (so reward *struct*s and classes and such, punish random wires)
- Numbers of interfaces to a module
- Module connectivity – how complex the connectivity graph is for a given IP.

To this point, these metrics have been applied to the IP or block level of design. The question now is:

- What about the complexity of an IP as a unit?
- How do we measure complexity at the SoC level?
- What techniques can we use to minimize this complexity?
- What overall benefits should we be able to realize by doing so?

M. Keating, *The Simple Art of SoC Design: Closing the Gap between RTL and ESL*, DOI 10.1007/978-1-4419-8586-6_9, © Synopsys, Inc. 2011

To answer these questions, we will extend some of the techniques used for design at the block level.

- We extend the concept of the command interface to the transaction interface, and develop a metric for the interface complexity of an IP
- We extend the concept of module connectivity to IP connectivity at the subsystem and then the SoC level

In looking at module connectivity we will use a concept similar to the cyclomatic complexity metric used in software.

Overview

Cyclomatic complexity is a method for measuring the complexity of software code. It was introduced by Thomas McCabe in his paper "A Complexity Measure" [17] published in 1976.

Cyclomatic complexity is a graph-based metric. It considers a piece of software (a function, method, class, etc) as a graph and calculates a metric based on the numbers of nodes and edges. The intent of the metric is to measure the number of linearly independent paths through the code.

Specifically, the nodes in the graph are blocks of statements that are always executed as a unit; that is, blocks of statements that have no conditional statements (*if*, *case*, *for*, etc).

The edges in the graph are directed edges; two nodes are connected if the second node (group of statements) might be executed immediately after the first node (group of statements). For instance, if there is an *if* statement in the first group of statements that (under a specific condition) calls the second group of statements, then there is a directed edge from the first group of statements to the second group.

Details

The cyclomatic complexity of a piece of code is the count of the number of linearly independent paths through the code. For instance, if the source code contains no decision points such as *if* statements, *case* statements or *for* loops, the complexity is 1, since there is only a single path through the code.

If the code has exactly one *if* statement containing exactly one condition there would be two paths through the code, one path where the *if* statement evaluates to *true* and one path where the *if* statement evaluates to *false*.

The formal mathematical equation for the cyclomatic complexity of a structured program (methods in object-oriented code) is defined on the control flow graph of the code to be:

$$CC = E - N + 2P$$

Where

CC is the cyclomatic complexity
E is the number of edges in the control flow graph
N is the number of vertices in the control flow graph

(continued)

P is defined as follows:

For a single method, P is always equal to 1. Cyclomatic complexity may, however, be applied to several methods at the same time (for example, to all of the methods in a class), and in these cases P will be equal to the number of methods in question, as each method will appear as a disconnected subset of the graph.

In his paper, McCabe shows that the cyclomatic number equals the maximum number of linearly independent paths in the code. That is, the cyclomatic complexity of any structured program with only one entry point and one exit point is equal to the number of decision points (conditional statements) contained in that program plus one.

For example:

```
public int getFoo (int bar) {
  int rslt = 0;
  if (bar == 0) {
    rslt = 2;
  } else {
    rslt = 0;
  }
  return rslt;
}
```

This code has two decision points (if and else) plus the entry point adds one, for a cyclomatic complexity of 3. Graphically:

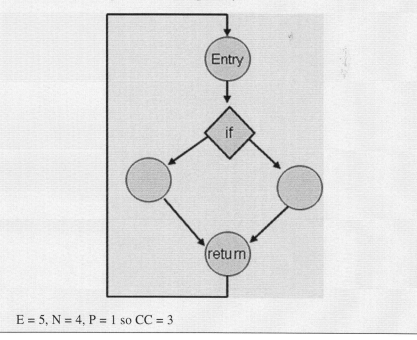

E = 5, N = 4, P = 1 so CC = 3

From Command to Transaction Interfaces

When talking about an IP block as a collection of modules, we described the interfaces as a set of commands (the control interface) and data. In most IP, the communication between modules is simple and the timing of this communication is relatively straight-forward. Command and data is typically delivered in one clock cycle, either directly from the sending module or from a FIFO.

When we look at an IP embedded in an SoC, the situation is different. Now, data and command may take several cycles to go from one IP (say, the CPU) over a bus to another IP (say, a data processing block like a JPEG core). The simple command/data style of interface is no longer a good model for this kind of communication.

Instead of looking at an IP as simply a collection of modules (as we did earlier in the book) we can look at it in the general form shown in Figure 9-1. This general model has data coming in and data coming out of the IP. The detailed behavior of the data path is controlled by the control registers, written through the bus interface from a CPU or host processor.

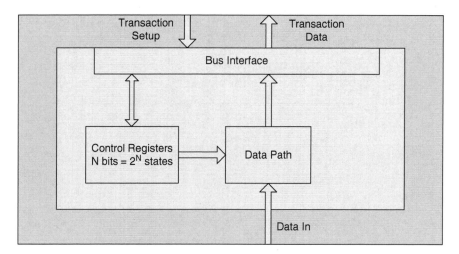

Figure 9-1 A general model for an IP.

We now think of communication to/from this IP as a set of transactions. Within the IP, a transaction may just consist of one module popping a control and/or data word off a FIFO that was written by a different module.

When we look at how the IP functions within the chip, transactions become more complex. Within the context of the chip, one transaction may consist of the CPU writing a large number of registers in the IP in order to set up the data path correctly. Another transaction may consist of the IP writing a large block of data over the bus to a on-chip memory.

There are several of ways of looking at these transactions, resulting in different kinds of metrics for the interface of the IP, and the complexity of IP as it behaves in the SoC.

- 1 register read/write = 1 transaction
- The complete setup of all control registers = 1 transaction
- The complete setup of all control registers + the transfer of data to memory = 1 transaction

JPEG Example

The JPEG IP (whose DCT we discussed in Chapter 7) has seven registers; they are nominally 32 bit registers, but only registers 1, 2, and 3 use any of the upper 16 bits. The register map is shown in Figure 9-2. In addition to setting up these registers, it is necessary to load quantization values and Huffman coding values into memory.

As we did in analyzing modules, we can separate these fields into data and control.

TABLE 5-1 JPEG CODEC Registers

Register	D15	D14	D13	D12	D11	D10	D9	D8	D7	D6	D5	D4	D3	D2	D1	D0
0															Start / Stop	
1	Not Used							Hdr	Ns		colspctype		De	Re	Nf	
2	NMCU [15:0]															
3	NRST															
4	H0				V0				NBLOCK0				QT0		HA0	HD0
5	H1				V1				NBLOCK1				QT1		HA1	HD1
6	H2				V2				NBLOCK2				QT2		HA2	HD2
7	H3				V3				NBLOCK3				QT3		HA3	HD3

Register	D31	D30	D29	D28	D27	D26	D25	D24	D23	D22	D21	D20	D19	D18	D17	D16
0	Not Used															
1	Ysiz (0–65,535)															
2	Not Used						NMCU [25:16]									
3	Xsiz (1–65,535)															

Figure 9-2   Registers 0-3 (32 bits) and Registers 4-7 (16 bits only).

Basically, registers 4-7 and the Huffman and Quantization tables are data: they affect the data path in the JPEG IP. Registers 0-3 affect the control of the JPEG:

- Register 0 starts and stops processing
- Register 1 determines whether we are encoding or decoding and how headers are processed

- Register 2 determines the number of Minimum Coded Units (MCUs) to process
- Register 3 determines the number of MCUs between start markers and the number of pixels per line

The fields NMCU, NRST, Ns, Ysiz and Xsiz are effectively deep state. The shallow state for these registers consists of eight bits:

- 1 bit from register 0
- 7 bits from register 1

Thus, the shallow state of the register set, as seen by the host CPU, is 2^8, or 256. This is a very manageable state space. Programming and operating the JPEG is a straight-forward process.

USB Example

The Wireless USB core (of which BCU, described earlier, is a part) is a much more complex design. It has approximately 127 registers, depending on the configuration. This is clearly a very large number of registers to read and write in order to set up correct operation of the core.

Many of the registers are effectively data: address pointers for where data should be delivered to/sent from by the core, and the like. But a very conservative estimate of the number of control bits in this register set is at least 142, so the shallow control state space of this register set is 2^{142}.

Clearly, this state space is way too large for a human to comprehend. Also clear is that there are not 2^{142} different operations that the core can perform. Most states in this state space are not useful – in essence, they are illegal states. But nothing in the register map makes it explicit which states are legal and which are not.

In order to manage this space – to make the core so that users can program it – we need to apply a layer of virtualization on it, to simplify it down to a human level. We need to create a layer above the core that gives the user a simple programming model.

Software Driver

The most common solution to this problem is to provide the software driver along with the RTL for the core. The software driver reduces the enormous register state space to a (relatively) small number of operations: reading and writing data, and managing the network. Effectively, the driver restricts usage of the core to those functions for which it was intended. It forces the user to program the register set only into configurations intended by the designers, and which have been tested.

An alternative solution to the problem of a huge register state space is to provide a layer of virtualization in hardware. Instead of seeing 127 registers, the programmer would see only one – which can be set to a small number of functions. These functions are then mapped by hardware into the (much larger number of) register reads/writes required to activate that function.

The advantage of the software driver as the virtualization layer is that it is flexible and does not impact the gate count of the core. The advantage of the hardware layer is that it is not flexible – the programmer cannot modify the mapping of function to registers, and so cannot put the core in a state not intended by the designers and not tested by them.

Virtual Platforms and Software Development

The key role played by the software driver is that it translates register reads and writes (the primitive transactions of an IP core connected to a bus) into a set of higher level transactions (such as reading a packet or processing a video frame). Having a trusted software driver for an IP can be an enabler for software development on virtual platforms.

Virtual platforms use behavioral models for the various components of an SoC to create a high level model of the chip for software developers to use while developing and debugging their embedded software. These models are too high level to model all the detailed behavior of the IP – at most they model registers for read/writes, but not the behavior that results from reading and writing these registers. So there is no way to verify that the specific reads and writes are correct. Instead, we rely on trusted software drivers to make sure that specific read and writes occur correctly. The virtual platform then models only the data transactions of the IP. By doing so, simulation on the virtual platform is extremely fast, allowing users to test and debug software effectively.

Connectivity and Complexity at the SoC Level

Given an IP core and its register map/software driver, we can calculate a complexity metric for it based on the ideas described above.

Once we have a collection of IP – and memory and IO – in the form of an SoC, we need a metric to calculate the complexity of the overall chip. And we will need techniques to keep this complexity under control.

There are several ways to look at a system. We start with a standard block diagram.

Function and Structure

The block diagram in Figure 9-3 shows a simple system.

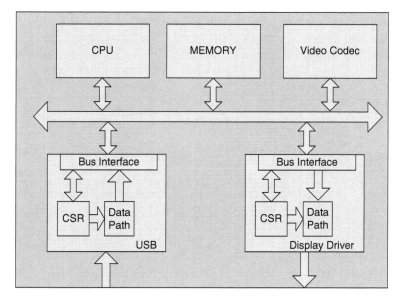

Figure 9-3 Diagram of a very simple SoC.

This block diagram shows the functionality of the system and the structure. A more complex system might consist of a number of subsystems; the block diagram can show this hierarchical structure.

In terms of complexity, the structure of the system, as shown in the block diagram, can tell us whether we have violated the rule of seven: if we have more than 7-9 blocks at any one level of hierarchy, then the structure is getting too complex, and we need to re-partition.

Connectivity and Bandwidth

We can view this system as a network or graph, as shown in Figure 9-4. In this graph, we can focus on which blocks can communicate with which blocks. In particular is shows what communication mechanism (in this case the Bus) is used for communication.

One of the major concerns in any modern SoC design is to assure adequate communication bandwidth between elements in the system.

In the trivial case shown in Figure 9-4, the bottlenecks are likely to be the bus and the shared memory. Assuming the video codec, USB and display driver all have DMA capability, there will be to be up to four bus masters competing for bandwidth through the bus to the memory.

Assuring adequate bandwidth early in the design cycle is challenging. We can come up with a reasonable guess of average bandwidth requirements, and try to design the bus and memory subsystem to provide at least this amount of bandwidth, and preferably more (to handle peak loads).

Typically, we can only achieve reasonable confidence in the communication/ memory bandwidth by running simulations of actual workloads with fairly accurate models. Early in the design of the system, we can use transaction level models – models of the interconnect and memory systems - to give us an indication of whether the bandwidth is sufficient. But the ultimate proof of performance is to run actual workloads on a detailed model of the design, such as an FPGA prototype or emulation system.

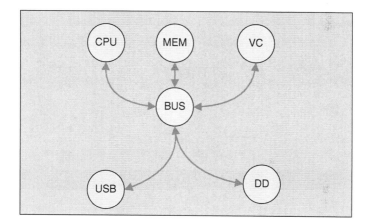

Figure 9-4 Connectivity graph for Figure 9-3.

Running real software on an FPGA prototype can also provide a large degree of confidence in the functional correctness of the hardware design. Because it can run many orders of magnitude faster than a simulator, the prototype can cover many more test cases than simulation. But there are still limits to what prototype testing can achieve, depending on the complexity of the system, and the size of the verification space.

Transactions and Complexity

One approach to assessing the overall complexity of a system is to analyze the transaction space of the design.

In the connectivity graph in Figure 9-5 we show the logical connections rather than the physical connections in the system. The bus is removed – since it only provides a path for transactions – and we show instead who can communicate with whom.

Note that in the block diagram we have 5 nodes (the 5 blocks). The bus structure implies that any block can communicate with any other block, resulting in a completely connected graph. Functionally, of course, this is not accurate, since (for example) the USB would never communicate directly to the display driver.

The Connectivity/Bandwidth graph shows the bus; but the bus does not generate transactions. So it affects the performance of the system, but not the functional complexity.

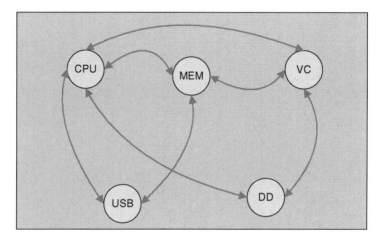

Figure 9-5 Graph showing the logical connectivity but not the physical connectivity for Figure 9-3.

The transaction graph in Figure 9-5 shows the possible paths of transactions. The CPU can talk directly to all the other blocks (to read/write registers and set up transactions, if nothing else). In addition, the USB and video codec communicate directly with memory, and the video codec communicates directly with the display driver.

Thus, we have 5 nodes and 7 edges in our graph. We then annotate the graph with the number of different transactions possible on each edge. For instance there may be exactly two possible transactions between the Video Codec and the Display Driver: the Video Codec can send a frame of data to the Display Driver, and the Display Driver can report status back to the Video codec.

The complexity of the system is then the product of all the possible transactions on all the edges.

If the design includes a lot of virtualization, as described earlier, then the number of transactions may be small: say, 5 transactions per edge. Then the complexity of the system is:

$$7^5 = 16,807$$

So every transaction could be tested, even in simulation. But even small sequences of transactions pose a challenge to simulation. Testing all possible two-transaction sequences requires

$$7^5 * 7^5 = \text{about 282 million transactions}$$

Assuming 20 clock cycles per transaction and a simulation speed of 100 cycles per second, this would require about 16,000 cpu hours. For an FPGA prototype running at 100MHz, though, this would take only about a minute. But even for an FPGA prototype, testing all possible sequences of 3 transactions would require:

$$7^5 * 7^5 * 7^5 = \text{about } 4.7 * 10^{12} \text{ transactions}$$

Or about 280 hours.

Limits to Chip Level Verification

These numbers indicate that only a very small subset of system functionality can be verified at the chip level – even if we use transaction level simulation or an FPGA prototype.

The first conclusion is that we cannot hope to verify IP at the chip level. We must build chips out of fully verified, trusted IP.

The second conclusion is that we can only do a limited amount of chip-level verification. We can verify that the basic transactions work and that the IP and buses are connected correctly. We can measure performance, and do some basic software testing.

Of course, we should do all we can do to minimize system complexity at the chip level. This will allow us to verify more of the system behavior and give us more confidence that it will function correctly once it is fabricated.

One strategy for minimizing complexity in a system as well as the verification effort is to manage the transaction complexity of the system. If we use virtualization (most likely through software) to restrict the possible sequences of transactions, we may be able to reduce the complexity of the system to a tractable level. The challenge is that there are a number of largely autonomous agents (in our case, the CPU, USB, and video codec) generating transactions, so there is no obvious way to assure exactly what sequences of transactions can occur.

But clearly, we need to manage system complexity by keeping the number of transaction types on any edge of the graph to a minimum. And we need to design and verify the chip at the transaction/software-driver level. If we deal with any blocks at the register read/write level, then we quickly (as in the case of the USB described earlier) end up with the number of transactions per edge in the range of 2^{100}, or essentially infinite.

Sub-systems and SoC Design

There are then two arguments for using hierarchy to partition a large SoC into a set of subsystems. The first, of course, is the rule of seven: it is difficult to reason about a system that has much more than 7 to 9 blocks at any one level of hierarchy. The second argument is that partitioning the system helps manage the transaction complexity of the system.

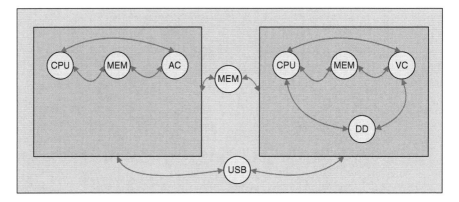

Figure 9-6 System partitioned into separate audio and video subsystems.

If we partition a system, say, into an audio and video subsystem (as shown in Figure 9-6), sharing a data input and a common memory, then we partition the transactions as well. The audio subsystem and video subsystem each may have a high number of internal transactions. But the transactions between each subsystem and the rest of the chip are limited to the raw data input and the processed audio/ video output.

SoC designs are so complex today that most are partitioned into multiple subsystems, or even hierarchies of subsystems (that is, subsystems of subsystems). Thus, at every level of design from the smallest module to the complete chip, we can use the basic strategy of encapsulation and well-designed interfaces to minimize the effective complexity of the system. We can structure the design so that global problems are broken up into a manageable set of smaller, local, tractable problems.

SoC Debug

One of the great challenges posed by an IP-based chip design methodology is that the IP and software can come from many sources. Some of the IP will come from third party providers and some will come from internal block developers. But with today's SoCs, the engineers designing and verifying the chip are not the engineers developing the IP. The result is that the chip designers do not know the internals of the IP. This makes debugging subtle problems at the chip level very difficult.

A similar problem is faced by the engineers writing the different layers of software that run on the chip. The engineers writing the drivers are not the engineers who wrote the RTOS. And the software engineers developing the application software typically do not know the internals of the RTOS or the drivers.

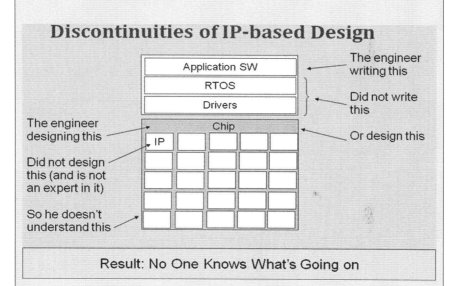

Although this IP-based SoC design methodology is essential for building complex SoCs, it can make debug very challenging. To constrain this challenge it is essential that the blocks are all robustly verified before integration into the chip.

It is also essential that the interfaces are well-designed and standardized. For hardware blocks this means standard bus interfaces such as AHB[20], AXI[20], and OCP[21]. It also helps if the transactions that occur between blocks are well defined and documented. For the chip design engineer, understanding the interfaces and transactions at the chip level are key for chip level debug.

(continued)

These complexities become even greater as the sources of IP expand. IP can be purchased from a third-party IP vendor, developed internally, or it can be legacy code from a previous project. It can be highly configurable, including configurable processors. Some of it may be generated from a high level synthesis tool, with the golden source code in C, C++, SystemC or Matlab.

The best way to integrate and verify IP from these various sources is still an open question. But having a high level model for the IP (and the chip) can allow verification engineers to develop and debug their tests more quickly and easily than on RTL.

One common technique is to use processor-driven testing. In this case, tests are written in C and run on the central processor. Bus functional models then drive the IO to/from the chip to make a complete testbench.

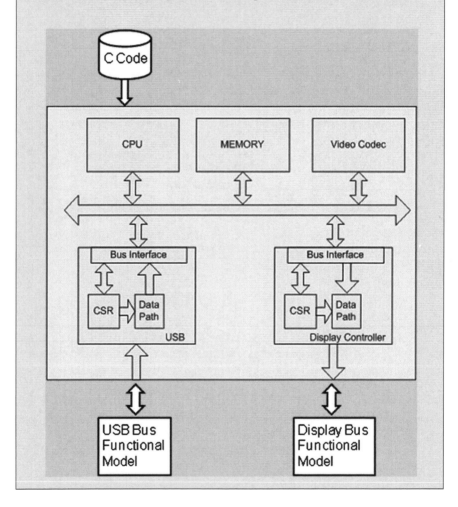

Summary

This chapter makes the following key points:

1. IP must be fully verified before it is integrated into the system. Very little functional verification can be performed at the chip level.
2. Sofware drivers are a key component of IP. Many IPs are simply too complex to be programmed correctly by software engineers not intimately aware of all the internal details of the design.
3. Hierarchy – that is, subsystems – are critical to managing the complexity of today's SoC.
4. At the subsystem and chip level, verification can and should focus only on verifying the connectivity and performance of the system, not its basic functionality.

One comment on this last point: there is a fundamental, underlying conflict between IP and SoC design. With today's practices and tools, it is not possible to completely verify a complex IP at an acceptable cost. We can make dramatic improvements from where we are today. And it is hoped that some of the ideas in this book will contribute to these improvements.

But no one has shown that it is possible to completely verify that a complex piece of code is functionally correct. It is not even clear if this is a well-defined problem – since the underlying intention of a design is often vague and poorly defined. The one hopeful sign is that the flight software of the space shuttle (about 500,000 LOC) appears to be bug free. But the cost of developing and testing this software was on the order of $1,000 per line of code. For a commercial IP of, say, 200,000 LOC, this would equate to a development cost of $200M. This cost would have to come down by at least a factor of 10 to be viable for chip and IP design.

So the intent of this chapter is to show how limited our ability is to verify at the chip and system level, and to give some guidance on how to structure designs for optimal verification at the chip level, and to emphasize the importance of bottom-up verification.

Chapter 10
Raising Abstraction Above RTL

In the previous chapters, we have discussed the challenges in SoC design and how to meet these challenges by using current tools and languages most effectively. In this chapter we extend our discussion to new paradigms, including high level design and synthesis.

The Challenge

As chip complexity rises faster than project budgets, there is a compelling need to improve designer productivity.

In the past, the move to RTL design (enabled by synthesis technology) and design reuse have provided significant boosts to designer productivity, particularly for SoC design. Raising the level of design above the RTL level is the most promising path to providing another significant boost.

To provide this benefit, raising the level of abstraction must address two problems:

1. It must enable designers to produce larger, more complex, but still optimal designs for the same effort.
2. It must dramatically reduce the amount of verification effort required for these complex designs.

The verification aspect of this problem is key. The two components of SoC design costs that are growing the fastest are functional verification and software. For many IP designs, the verification effort is 80% or more of the total project effort. Clearly, no significant progress can be made in improving overall design productivity without addressing the verification challenge.

M. Keating, *The Simple Art of SoC Design: Closing the Gap between RTL and ESL*,
DOI 10.1007/978-1-4419-8586-6_10, © Synopsys, Inc. 2011

High level design and high level synthesis provide a promising approach to these problems. The basic objective of current high level synthesis tools is to allow the designer to code at the untimed, algorithmic level, typically in C/C++/SystemC. Automated synthesis then converts the untimed model into a fully timed, pipelined model, typically in Verilog, that can be synthesized into gates by traditional synthesis tools.

The advantages of designing and coding at this level are:

1. The designer can evaluate many different architectures before committing to detailed design.
2. The amount of code required for a given design decreases by an order of magnitude, typically reducing the number of bugs by a similar amount.
3. Untimed code is much easier to test, debug, and reason about.

Current High Level Synthesis Tools

The current high level synthesis tools have had some success in some designs. In particular, these tools do well with algorithmic designs such as video and audio codecs, where loops of complex arithmetic operations dominate the design. The tools can partially unroll the loops, schedule the operations, optimize the sharing of multipliers, and so on.

For example, consider the DCT from Chapter 7. A fragment of the C code for this looks like Example 10-1:

```
for(y=0;y<8;y++){
  for(x=0;x<8;x++){
    for(u=0;u<8;u++){
      v=block[y][x]*c[x][u];
      v+=2048;
      v>>=12;
      if(s[x][u]) v=-v;
      reg[u]+=v;
    }
  }
}
```

Example 10-1

The high level synthesis tool maps this to a hardware template. One such hardware template is shown in Figure 10-1. The loops of arithmetic operations are mapped onto a set of hardware resources such as multipliers and adders (or an ALU), with scratch pad memories and registers to hold temporary values. These resources are shown on the right side of Figure 10-1.

The looping itself is controlled by a state machine, shown on the left side of Figure 10-1.

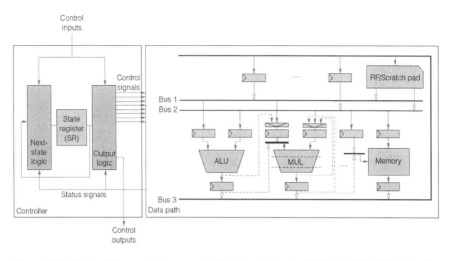

Figure 10-1 Typical Target Architecture for High Level Synthesis[22]. © 2009 IEEE. Used by Permission.

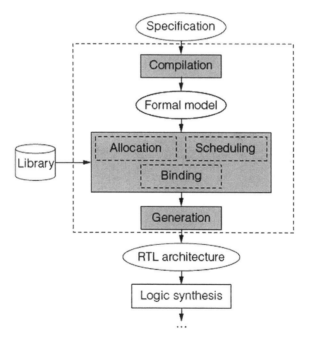

Figure 10-2 Basic High Level Synthesis Design Steps[22]. © 2009 IEEE. Used by Permission.

The high level synthesis process is shown in Figure 10-2. The key steps in this process are scheduling, allocation and binding. These steps determine which arithmetic operations are executed in which cycle, how operators are shared, and how the resources are used. Figure 10-3 shows these steps in more detail: the design is mapped to a graph of operations, and then the graph is partitioned (scheduled) into clock cycles. Finally, this schedule is mapped onto hardware resources like ALUs and registers.

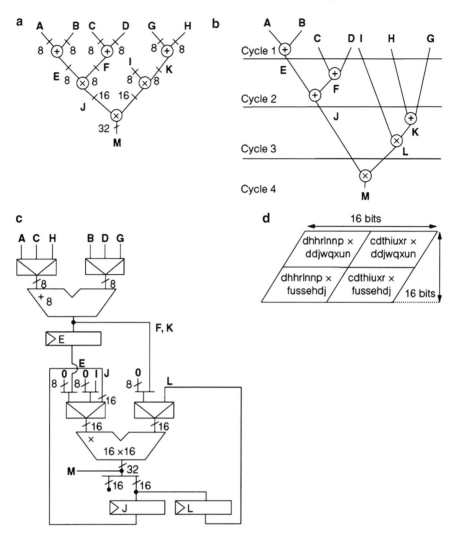

Figure 10-3 Example of Scheduling, Binding and Allocation[23]. © 2009 IEEE. Used by Permission.

Typically, code such as that shown in Example 10-1 will not produce acceptable QOR (in particular area) if synthesized directly. The designer usually must either modify the code or provide guidance to the tool (through a script or GUI). At the very least, the designer needs to tell the tool what the target latency and cycle time are. Often, the designer needs to guide the tool in terms of which loops to unroll and which loops not to unroll.

In addition, typical hardware designs operate on variables that may not be an "int" in C. The DCT, for example, may operate on 8-bit or 16-bit pixels. For area efficiency, we must indicate to the synthesis tool, either in code or via the GUI, the correct size of the variables. Tools typically use some form of extended C (often SystemC data types) for specifying variable sizes.

Thus, the C code that is synthesized is not the purely untimed, integer C we started out with, but rather C code to which considerable timing and sizing information has been added. Most high level synthesis tools provide some capability for simulating this partially timed C code before synthesis.

For some designs, like the DCT, providing timing information is not sufficient to produce good QOR from high level synthesis. There are patterns in the constant arrays (c[x][u] in the code fragment in Example 10-1) that can be exploited to make the arithmetic much more efficient. Most current tools cannot spot and exploit these patterns; rather, the C code must be modified to produce a design that is comparable to hand-coded RTL in area.

There are many advantages in using high level design and synthesis for DSP-type applications. The amount of code can be reduced significantly: in some cases it can be reduced by a factor of 10x. There are typically fewer bugs (if only because there are fewer lines of code), and these bugs are typically found earlier in the design cycle.

Successes

Users of C-level synthesis have reported significant successes, especially in datapath-centric designs like video and audio codecs. One design team has reported doing a complete video codec that supports multiple standards including H.264. This design was coded entirely in ANSI C. Compiling this C code with the High Level Synthesis tool produced a single hierarchical RTL design. No manual intervention was required to generate production-ready RTL. The resulting RTL synthesizes to about 700k gates.

Based on this experience, and previous projects using high level synthesis, the design team reports a code compaction of 3x over RTL. The combination of fewer lines of code and a C-based verification methodology has had a significant impact on verification.

C-Level Verification

Verifying the C code was significantly easier than for an equivalent RTL design. The team reported a 3x reduction in bugs as compared to RTL coding. The fact that C has a sequential programming model (that is, there is no concurrency in the C code) makes the C code fundamentally easier to debug than RTL. Debugging an issue in C took half the time taken to debug an issue in RTL. Due to the speed of C simulation the team was able to verify the algorithmic design much faster than in RTL.

RTL Verification

The C code was extensively verified by the design team before synthesis. After synthesis, the RTL was verified by a separate verification team. This

(continued)

team reported that the synthesized RTL was more thoroughly verified – and had significantly fewer bugs – than the typical hand-coded RTL that they receive.

After an initial cycle of debug, the RTL verification became more of a regression test as the C design was refined and re-synthesized.

One lesson learned was that the more code that is encapsulated in C, the better. When hand coded and synthesized RTL are combined, bugs become harder to find and verification becomes more difficult.

Bottom Line

The overall verification effort for the project was estimated to be about half that of verifying an equivalent RTL design created by hand. This resulted in a significant productivity improvement and reduction in the overall project cycle time.

But there are still some significant opportunities for improving the current versions of these tools:

1. It often takes a significant effort by the designer to produce an untimed model that is competitive in area with a hand-coded design.
2. Users have found that a significant amount of verification must still be done at the RTL level. The successful results mentioned above achieve a 2x overall improvement, but we need a path to still larger gains.
3. Most users find the current tools inadequate for control-dominated designs.

Part of the problem is that the scheduling and pipelining algorithms used by these tools need to improve, and over time they certainly will.

Another part of the problems is that there is a huge gap between high level design (as supported by today's tools) and RTL design. Specifically:

	High Level Design	RTL Design
Abstraction	Untimed	Fully timed, pipelined
Language	C++/SystemC	Verilog/SystemVerilog
Functional Debug	Source code debug (gdb)	Waveform Viewer

So the two approaches are different in abstraction, language and tools. This is a very large gap. Imagine if gate level netlists were in a different language from RTL, and required an entirely different set of simulation tools.

As long as such a dramatic gap exists, high level design will be very limited in its adoption and in the value it can deliver to designers. The goal of this chapter is to discuss how to close the gap between high level design and RTL design.

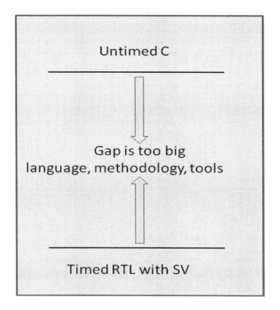

Figure 10-4 Gap between RTL and high level synthesis is too large.

Closing the Abstraction Gap

The fundamental problem with the current approach of C-based high level synthesis is its premise that it is possible to synthesize high-performance RTL from untimed C. The reality is that synthesis requires a "partially timed" design description that contains more low level information than the pure C code. At the same time, unless a full RTL language is used (such as SystemC), it is not possible to describe the detailed timing relationships required for many control-dominated designs, such as a USB or DRAM interface.

On the other hand, typical RTL contains a great deal of low level information that is redundant and unnecessary. The first few chapters of this book describe the "syntactic fluff" and unstructured code found in most designs. More fundamentally, RTL code (and RTL synthesis tools) requires a detailed, cycle-by-cycle description of the circuit.

In many cases this is more information than the designer cares about. For instance, the design specification may allow a range of latencies; it may allow a tradeoff between latency and power. The specification may also allow a slower or faster start-up sequence than the RTL describes. Or the RTL may describe a precise timing between when registers are written and when they affect the behavior of a design, even if the designer knows that registers are only written when the circuit is not active.

Thus, the detail required for RTL demands that a design over-specify the behavior of a design. There are no timing "don't cares" in RTL the way there are logical "don't cares."

To improve the productivity of hardware design and verification, to raise the abstraction of design across the whole spectrum from data path to control code, we need to find a flexible way of describing circuit behavior that has the right amount of detail for the problem being solved. That is, we need code that contains the optimum amount of (timing) information.

Designers need to provide enough information so that tools can do a good job of scheduling and optimization. But designers should not have to provide any additional information. The less timing (and variable sizing) information in the code, the easier the code is to review manually and to verify automatically.

But above all else, we need to provide a productive design environment where we can move seamlessly from very high levels of abstraction (untimed C) to very low levels of abstraction (RTL SystemVerilog) as required to solve a specific design problem.

SystemC

High Level Synthesis tool providers have recognized that untimed C does not address the needs of designers. Most have extended their tools to support SystemC – a template library (and reference simulator) that extends C to be a full RTL language. Unfortunately, SystemC does not really address the problems outlined above. It is a template library added to a very large language (C++), not a domain specific hardware description language (see section on Domain Specific Languages below).

SystemC has proven to be very useful as a high level modeling language for virtual platforms. It allows software developers to work in a C++ environment for developing and debugging their embedded software. Because users are debugging their software, and not the SystemC models, the shortcomings of SystemC are not a major impediment.

But as a hardware design language, SystemC has major liabilities. It is verbose, cumbersome to use and very difficult to debug. And it has much more syntactic fluff than Verilog.

A current trend is to extend C-based high level synthesis with SystemC to provide the hardware-specific capabilities needed for high level design. Unfortunately, because of SystemC's shortcomings, this combination has not yielded the productivity gains designers are seeking.

The Right Usage Model for High Level Design

The adoption of SystemC as a hardware design language is an accident of history. Most high level synthesis tools were developed before SystemVerilog was developed. Verilog95 and Verilog2001 offer little support for abstract design. But most importantly, initial marketing of high level synthesis completely misunderstood the market. HLS companies initially envisioned that algorithm developers could use HLS tools to design hardware. Over the last half dozen years or so, it has become clear that this is not going to happen.

Algorithm development and hardware design are two separate disciplines requiring different skill sets and knowledge. Good design requires that architects, algorithm developers and hardware designers work together. Forcing hardware designers to work in C or SystemC (neither of which is acceptable as a hardware design language) makes no more sense than forcing algorithm developers to use Verilog95.

The solution is to have an environment that allows algorithm developers and hardware designers to work together effectively. Such an environment must support C++, must allow C++ to be extended with hardware data types (arbitrary sized variables), and must provide a full RTL capability.

Extending C++ to become a true HDL would create yet another language, and this seems an unlikely and unnecessary path. It is much more feasible to use SystemVerilog and C++ together to achieve our goals.

SystemVerilog as a High Level Design Language

SystemVerilog already employs many C/C++ constructs (classes, structs, enumerated types). It also supports RTL as well as extensive hardware verification capabilities. Also, most simulators support C/SystemVerilog co-simulation. Thus, combining C and SystemVerilog seems the right basis for moving forward.

Example 10-2 shows (part of) the C code for the DCT on the left. On the right it shows the same code converted to SystemVerilog. The changes required are:

1) Adding apostrophes to the constant array declarations (*int c*)
2) Changing *void dct* to *function* dct (and adding *endfunction*)
3) Converting *{}* to *begin end*
4) Changing the variable *reg* to *reg_x* since *reg* is a reserved work in Verilog

These changes are very minor, and could be automated. Thus, with minor modifications, a SystemVerilog simulator could be enhanced to support this level of C natively. Then it would be possible to take the original C code and incrementally modify variables from integers to SystemVerilog types such as *bit* and *logic*, while remaining in the same environment and using the same debugger. This would give C code all the data types required for high level synthesis, while providing the full capability of SystemVerilog to describe lower level aspects of the design.

Thus, combining C and SystemVerilog would allow a smooth "lowering" of C to start closing the gap between untimed C and RTL design abstractions.

```
int block[8][8];                      int block[8][8];

int s[8][8]={                         int s[8][8]='{
{0, 0, 0, 0, 0, 0, 0, 0},             '{0, 0, 0, 0, 0, 0, 0, 0},
{0, 0, 0, 1, 1, 1, 1, 1},             '{0, 0, 0, 1, 1, 1, 1, 1},
{0, 0, 1, 1, 1, 0, 0, 0},             '{0, 0, 1, 1, 1, 0, 0, 0},
{0, 0, 1, 1, 0, 0, 1, 1},             '{0, 0, 1, 1, 0, 0, 1, 1},
{0, 1, 1, 0, 0, 1, 1, 0},             '{0, 1, 1, 0, 0, 1, 1, 0},
{0, 1, 1, 0, 1, 1, 0, 1},             '{0, 1, 1, 0, 1, 1, 0, 1},
{0, 1, 0, 0, 1, 0, 1, 0},             '{0, 1, 0, 0, 1, 0, 1, 0},
{0, 1, 0, 1, 0, 1, 0, 1}              '{0, 1, 0, 1, 0, 1, 0, 1}
};                                          };


int pixout;                           int pixout;

void dct(){                           function dct();
  int y,x,u,v;                          int y,x,u,v;
  int reg[8];                           int reg_x[8];

/* Horizontal */                      /* Horizontal */

for(y=0;y<8;y++){                      for(y=0;y<8;y++) begin
  for(x=0;x<8;x++)                       for(x=0;x<8;x++)
    reg[x]=0;                             reg_x[x]=0;

  for(x=0;x<8;x++){                      for(x=0;x<8;x++) begin
    for(u=0;u<8;u++){                      for(u=0;u<8;u++) begin
      v=block[y][x]*c[x][u];               v=block[y][x]*c[x][u];
      v+=2048;                             v+=2048;
      v>>=12;                              v>>=12;
      if(s[x][u]) v=-v;                     if(s[x][u]) v=-v;
      reg[u]+=v;                           reg_x[u]+=v;
    }                                     end
  }                                     end

  for(x=0;x<8;x++) {                     for(x=0;x<8;x++) begin
      block[y][x]=reg[x];                  block[y][x]=reg_x[x];
  }                                     end
}                                     end
                                      endfunction
```

Example 10-2 Left = C, Right = SystemVerilog. Changes are highlighted in red.

But to make this work, we need to raise the level of abstraction of RTL, specifi-cally the synthesizable subset of SystemVerilog. SystemVerilog is the best infra-structure on which to build a true high-level design environment. But as shown in the earlier chapters of this book, SystemVerilog as it is typically used is too low level for complex designs.

Raising the Level of Abstraction of RTL

At first, it might seem improbable that we can raise the level of abstraction while remaining at the RTL level. But the previous chapters show that there is a huge difference between well-written and poorly-written RTL, in terms of clarity of code and design complexity.

A number of authors have remarked that C is just portable assembly language. But this misses a key point. Although they operate at approximately the same level of abstraction, C provides a means of expressing this abstraction that is much more powerful than assembly language. In particular, it allows the development of pro-grams that are much larger, more structured, and easier to get right than assembly language.

C++, of course, provides a whole new paradigm in programming – object oriented software – which provides a dramatic increase in abstraction above C. But C provides significant, incremental improvements of abstraction over assembly language.

Perhaps it would be useful to try to define what we mean by raising the level of abstraction. At the very least, we mean enabling code that is more concise, easier to understand, and more natural for the problem it is solving. There are two basic approaches to raising the abstraction of design:

1) Leaving out (implementation) details that can be addressed by a compiler
2) Using more appropriate primitives, especially primitives that express a design in a simple, appropriate way

C-synthesis tools use the first approach, leaving out timing details. But they are playing catch-up in the second approach, since they have eliminated many of the useful primitives (logic/bit) of Verilog, and are forced to extend C to put them back.

Matlab uses the second approach. By providing matrices and matrix arithmetic as primitives in the language, complex algorithms can be expressed (and tested) at a very high level of abstraction. The equivalent code in C or Verilog would require multiple for-loops, greatly obscuring the underlying intent of the code. Note that in focusing on the second approach, Matlab also addresses the first approach – using higher level primitives often results in leaving out implementation details. Matlab is a good example of a domain specific language.

The proposals in this book use both the first and the second approach to raise the level of abstraction of RTL above the level of Verilog (as it is commonly used today). The first approach is represented in recommendations not to "pre-synthe-size" code. For instance, we should not code a multiply by constant as a shift and add. Rather, we should leave it as a multiply, and let synthesis do the optimization. By leaving out such low-level implementation details we produce cleaner code and, it turns out, better synthesis results.

The recommendations to use tasks, functions, and a single state machine per module all represent a shift of which SystemVerilog primitives should be used to code RTL. By using these more appropriate, higher level constructs we raise the level of abstraction in our RTL code.

Domain Specific Languages

Domain-specific languages are a hot topic in software. One of the challenges in software development today is that both the applications (like speech recognition or internet searches) and the software development environment (C++ on parallelizing compilers targeting multi-core hardware) are getting extremely complex. To create the best software, a programmer has to be a world-class expert in the problem domain and in software development. This combination is becoming increasingly difficult to achieve.

One approach is to change the programming environment, to modify the programming language to allow the developer to deal with constructs and operations that are specific and natural to the problem domain.

Martin Fowler provides extensive information on this topic in his book *Domain Specific Languages*[13].

SystemVerilog as a Domain Specific Language

Hardware design requires a domain specific language to meet the needs of today's very complex designs. Unfortunately, SystemVerilog does not provide such a capability.

Just as C++ facilitates a higher level of abstraction than C, SystemVerilog provides incremental improvements over Verilog. In its verification capabilities, it provides a host of primitives not available in Verilog, including classes, dynamic and associative arrays, mailboxes, queues, assertions (including sequences) and many more.

On the hardware description side of SystemVerilog (the synthesizable subset), enumerated types, structs and unions, and interfaces provide abstraction mechanisms that allow the engineer to code synthesizable RTL that is more structured than Verilog.

But there are still some fundamental shortcomings in the synthesizable subset of SystemVerilog. Largely for historical reasons, SystemVerilog does not provide constructs specific to and appropriate for the domain of digital hardware design. Verilog was developed before synthesis tools were available. It was intended primarily as a general-purpose (hardware) simulation language. So "always" is really

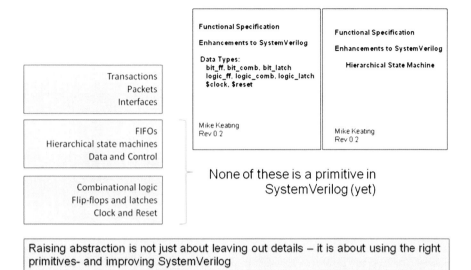

Figure 10-5 Defining the right primitives to make SystemVerilog a domain specific language for design.

a simulation directive; later, it was used by synthesis to indicate a flop ("always @ (posedge clk ...)") or combination logic ("always @ (sig1 or sig2 ...)). Now that synthesis is such a fundamental design tool, we need to revisit the hardware description side of SystemVerilog and make it a much more domain-specific language – that is, raise its level of abstraction.

SystemVerilog Primitives

Designers of digital hardware think in terms of the following primitives.

- Combinational gates
- Flip-flops
- Latches
- Clock
- Reset

Above this level, designers think in terms of:

- FIFOs
- Finite State Machines

Above this level, designers think in terms of :

- Packets
- Transactions
- Operations on matrices of pixels

Note that none of these are primitives (types) in SystemVerilog. Instead we have

- Bit and logic
- Wire
- Reg

Whether a specific reg (or wire or bit) is a flip flop or a logic gate depends on how it is used (always_ff or always_comb, etc.) It makes much more sense to define a variable to be of type combinational, flip-flop, clock, and so on. This would allow more effective automatic checking of the code, as well as eliminate a lot of syntactic fluff that obscures RTL code today.

Proposal

We propose updating the synthesizable subset of SystemVerilog to make it a true domain specific language for digital hardware design. Doing so will shrink the gap between C-level design and RTL design, and provide a path to addressing the major

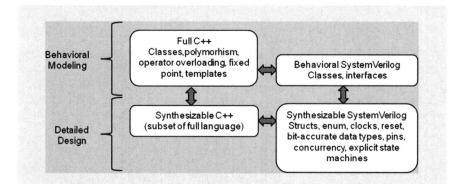

C++ and SystemVerilog provide the key capabilities required for hardware design. At the behavioral level, C++ offers a very rich set of features, including polymorphism, operator overloading, and so on. C++ allows developers to describe very complex systems, without worrying about what will end up in hardware and what will end up as software. SystemVerilog overlaps C++ as a behavioral language, providing classes, interfaces, and other object-oriented features. But SystemVerilog is not as rich or as general as C++.

SystemVerilog has many specialized features for hardware design, including clocks, resets, state machines, as well as higher level features such as structs and enums. Synthesizable C++ overlaps SystemVerilog as a detailed design language, but is not as rich or as general as SystemVerilog.

The ideal design environment enables engineering teams to work in both languages, smoothly transitioning from one to the other as required for a specific design task.

short-coming of high level synthesis – how to raise the level of abstraction in control-dominated design.

The next chapter presents some initial ideas of what this extended SystemVerilog should look like. Eventually, it should support all of the primitives listed above as first-class objects in the language: combinational gates, flip-flops, latches, clock, reset, FIFOs, finite state machines, packets, transactions, and matrices. But we confine our initial focus to the finite state machine and the primitives needed to support it: combinational gates, flip-flops, latches, clock, and reset.

Chapter 11
SystemVerilog Extensions

Overview

The extensions described in this chapter focus on providing an efficient finite state machine primitive for SystemVerilog. In earlier chapters, we described the challenges of state machine design and the value of reducing state space to a minimum. Using a hierarchical finite state machine can significantly reduce the complexity of a design, often by several orders of magnitude. But there is no explicit support in SystemVerilog for FSMs in general or for hierarchical FSM specifically. Thus, there is no uniform way of coding them.

Providing a uniform coding mechanism for a hierarchical finite state machine provides many benefits to the designer and to the EDA tools that assist the designer:

1) All state machines would have the same fundamental structure, making design review much easier
2) State machines would be explicit, enabling tools to do more checking and analysis, as well as more optimized simulation and synthesis.
3) Hierarchical state machines would be easier to code, encouraging designers to use them.

Basic Extensions

We start with Example 11-1. It shows a simple state machine that reads a packet from the input and sends the packet to the output.

The key changes in the code from standard SystemVerilog are the use of :

1. *smodule*
2. *bit_ff*

M. Keating, *The Simple Art of SoC Design: Closing the Gap between RTL and ESL*, DOI 10.1007/978-1-4419-8586-6_11, © Synopsys, Inc. 2011

3. *bit_comb*
4. *$clock*
5. *$reset*
6. *top_state_machine, sub_state_machine*
7. *state_var*
8. *done* (to signal the end of a sub-state machine activity)
9. combinational and sequential assignments in the same case statement

Each of these features is described in the next section.

```
smodule foo (
  input bit clk, resetn,
  input bit pkt_in_fifo_empty, pkt_out_fifo_full,
  input bit [31:0] data_in,
  output bit_ff [7:0] data_out = 8'h6,
  output bit_comb got_pkt,
  output bit_comb in_pkt_pop ) ;

  $clock posedge (clk) ;
  $reset async negedge (resetn) ;

  struct packed{
    bit [7:0] header;
    bit [7:0] destination;
    bit [7:0] payload;
    bit [7:0] crc; } input_packet;
//-------------------- main state machine ------
state_machine tctrl () ;

  typedef state_var {IDLE,GET_PKT,SEND_MPKT}
    tx_state_type;

  tx_state_type tx_state = GET_PKT;

  case (tx_state)
   IDLE:if (!pkt_in_fifo_empty)
        tx_state <= GET_PKT;
   GET_PKT: begin
    get_input_packet();
    if (get_input_packet.done)
        tx_state <= SEND_MPKT;
   end
```

Example 11-1

```
  SEND_MPKT: begin
   send_mpkt_sm();
   if (send_mpkt_sm.done)
       tx_state <= GET_PKT;
  end
  default: ;
 endcase
endstate_machine //tctrl

//--------------gpkt sub-state machine ---------
state_machine get_input_packet();
 state_var {GET_PACKET,DONE}gpkt_state=GET_PACKET;
 done =0;
 case (gpkt_state)
  GET_PACKET: begin
   input_packet.header <= data_in[31:24];
   input_packet.destination <= data_in[23:16];
   input_packet.payload <= data_in[15:8];
   input_packet.crc <= data_in[7:0];
   in_pkt_pop <= 1;
   got_pkt = 1;
   gpkt_state <= DONE;
  end
  DONE: begin
   done = 1;
   got_pkt = 0;
   in_pkt_pop <= 0;
   gpkt_state <= GET_PACKET;
  end
  default:;
 endcase
endstate_machine

//------------- mpkt state machine ----------
state_machine send_mpkt_sm();
 state_var {
  SEND_HEADER,SEND_DEST,SEND_PAYLOAD,SEND_CRC }
   mpkt_state = SEND_HEADER;

 done =0;
 case (mpkt_state)
  SEND_HEADER: begin
```

Example 11-1 (continued)

```
     if (! pkt_out_fifo_full) begin
       data_out <= input_packet.header;
       mpkt_state <= SEND_DEST;
     end
   end
   SEND_DEST: begin
     if (! pkt_out_fifo_full) begin
       data_out <= input_packet.destination;
       mpkt_state <= SEND_PAYLOAD;
     end
   end
   SEND_PAYLOAD: begin
     if (! pkt_out_fifo_full) begin
       data_out <= input_packet.payload;
       mpkt_state <= SEND_CRC;
     end
   end
   SEND_CRC: begin
     if (! pkt_out_fifo_full) begin
       data_out <= input_packet.crc;
       mpkt_state <= SEND_HEADER;
       done = 1;
     end
   end
   default: ;
  endcase
 endstate_machine

 tctrl ();// = always_ff@(posedge clk) tctrl ();

 endmodule
```

Example 11-1 (continued)

smodule

We define a special kind of module (*smodule*, or synchronous module) where we can make certain assumptions about the design that are not true of all designs (and hence not true of modules in general). In an *smodule* we assume (and require):

1. There is only one clock for the module. All flops are clocked on a same edge of this clock, either *posedge* or *negedge*.
2. There is only one reset for the module.

   Within an *smodule*, the features described below can be used.

bit_ff

In an *smodule*, we declare signals to be of type *bit_ff* (or *logic_ff*) if they are flops, *bit_latch* (or *logic_latch*) if they are latches. Any signal declared to be of type *bit_ff* will be clocked on each clock cycle, using the clock (and edge) defined in the $clock statement.

Assignments to a *bit_ff* variable must use the <= (non-blocking) assignment operator.

Signals of type *bit_ff* may have an initialization value, which is used for its reset value. Every *bit_ff* which is declared to have an initialization value will be reset using the reset signal specified by $reset.

Examples:

```
output bit_ff [7:0] data_out = 8'h6;
bit_ff [31:0] foo = 32'hdeadbeef;
```

bit_comb

In an *smodule*, we declare signals to be of type *bit_comb* (or *logic_comb*) if they are combinational signals.

Assignments to a *bit_comb* variable must use the = (blocking) assignment operator.

Examples:

```
output bit_comb in_pkt_pop;
bit_comb[17:0] bar;
```

$clock

$clock defines which edge of which input signal is used for the clock for the *smodule*. Only one clock can be defined, and it will be used for all flops.

$reset

$reset defines which edge of which input signal is used for the reset signal for the *smodule*, and whether it is synchronous or asynchronous. Only one reset can be defined, and it will be used for all flops that have an initialization value defined.

state_machine

The features described above allow us to define a very concise construct for the hierarchical state machine. The top state machine (**state_machine** tctrl) defines the main state machine. We instantiate it explicitly at the top level of the module (see the next to last line of Example 11-1). As a result, it is treated like a task that is called every clock cycle (that is, on the clock edge defined by *$clock*).

The top state machine can call sub state machines (**state_machine** get_ input_packet). Sub state machines are like tasks that are called when they are invoked by another state machine (either the top state machine or another sub state machine).

state_var

The state machine uses a special kind of enum statement to define the state variable for the state machine (***state_var** {GET_PACKET, DONE} gpkt_state = GET_ PACKET;*). This *state_var* defines the states for the state machine and the reset state. We use the *state_var* rather than a regular *enum* because it enables the simulation and synthesis tools to do analysis and optimizations based on the fact that it is a state variable for a state machine.

done (to signal the end of a sub-state machine activity)

Each sub state machine has a predefined Boolean signal called *done*. It is used by the sub state machine to signal to the calling state machine that the sub state machine is done, and that control goes back to the calling state machine.

combinational and sequential assignments in the same case statement

A key feature of the state machine construct is the ability to mix combinational and sequential assignments in the body (case statement) of the state machine. This is enabled by the *bit_ff/bit_comb* features that define explicitly how these assignments are to be interpreted. In standard Verilog, sequential code and combinational code would have to be in separate processes, since the process type (always_comb or always_ff) would be the only indication of whether the assignment was a combinational assignment or a sequential assignment.

With the *bit_ff/bit_comb* types we can mix both types of assignments in the same case statement. We know that a variable of type *bit_ff* will synthesize to a flip-flop, and that in simulation the assignment will occur at the next clock. We know that a variable of type *bit_comb* will synthesize to logic gates, and that in simulation the assignment will occur immediately.

The result is that all of the actions in a particular state (both combinational and sequential assignments) are co-located in the same section of code, making it easier to review and understand the code.

Other Capabilities

There are other capabilities for *smodules* not shown Example 11-1. They are described in the next sections.

First

The state machine has a pre-defined function *first* that returns a one if it is the first clock cycle that the state machine is in that state, and a zero otherwise. For example:

```
case (mpkt_state)
  SEND_HEADER: begin
    if (first())
      $display("first cycle in SEND_HEADER");
    if (! pkt_out_fifo_full) begin
      data_out <= input_packet.header;
      mpkt_state <= SEND_DEST;
    end
  end
```

Functions and Tasks

As in regular modules, functions and tasks can be used to structure code. Functions and tasks can be called by state machines.

Assignments Outside State Machines

Because of the capabilities provided by *bit_ff/bit_comb/$clock/$reset*, assignments in *smodules* can be made without using *always* statements, as shown in Example 11-2.

In this code:

1. *header* is defined as a flop, which will be reset to all ones
2. *payload* is defined as a flop, which will be reset to zero
3. The assignment to header is a non-blocking assignment to a *bit_ff*; no *always_ff* is required.
4. *out_comb* is defined as a combinational signal. It is assigned using a blocking operator. No *always_comb* is required.

```
smodule foo (
   input bit clk, resetn,
   input bit pkt_in_fifo_empty, out_comb_control,
   input bit [31:0] data_in,
   output bit_comb [31:0] out_comb
   ) ;

   $clock posedge (clk) ;
   $reset async negedge (resetn) ;

   bit_ff [15:0] header = 16'hffff;
   bit_ff [15:0] payload = 16'h0;

   if (!pkt_in_fifo_empty) begin
      header <= data_in[31:16];
      payload <= data_in[15:0];
   end

   if (out_comb_control) out_comb = {header, payload};
   else out_comb = 32'b0;
endmodule
```

Example 11-2

Iterative State Loops

In some state machines, it is convenient to iterate over a state variable. For this we introduce the *for-istate* construct.

```
case (spkt_state)
   SEND_HEADER: begin
      for (istate i = 0; i < 8; i++) begin
          bit_stream_out <= input_packet.header[i];
      end
      spkt_state <= SEND_DEST;
   end
```

Example 11-3

In Example 11-3, we declare the loop variable (i) to be of type *istate*. As a result, each iteration through the loop is treated as a separate state. We output eight bits, one per clock cycle. Compare this to Example 11-4.

```
case (spkt_state)
   SEND_HEADER: begin
      for (int i = 0; i < 8; i++) begin
          stream_out <= input_packet.header[i];
      end
      spkt_state <= SEND_DEST;
   end
```

Example 11-4

In example 11-4, the loop variable is declared as type *int*. In this case, all 8 bits are output at the same time; the synthesis tool unrolls the *for* loop and all 8 sequential assignments happen on the same clock edge.

We extend SystemVerilog, then, to include an iteration mechanism (*for istate*) which indicates that each iteration through the loop corresponds to a state in a state machine. By declaring the loop variable to be of type *istate* (iterative state), we specify an inferred state machine with the appropriate number of states. So the *istate* in Example 11-3 is equivalent to Example 11-5.

```
case (spkt_state)
  SEND_HEADER: begin
    case (i)
      0: begin
        bit_stream_out <= input_packet.header[i];
        i++;
      end
      1: begin
        bit_stream_out <= input_packet.header[1];
        i++;
      end
      2: begin
        bit_stream_out <= input_packet.header[1];
        i++;
      end
      3: begin
        bit_stream_out <= input_packet.header[1];
        i++;
      end
      4: begin
        bit_stream_out <= input_packet.header[1];
        i++;
      end
      5: begin
        bit_stream_out <= input_packet.header[1];
        i++;
      end
      6: begin
        bit_stream_out <= input_packet.header[1];
        i++;
      end
      7: begin
        bit_stream_out <= input_packet.header[i];
        i = 0;
      end
    spkt_state <= SEND_DEST;
  end
```

Example 11-5

This *istate* iteration can be used in DSP applications, like the DCT, to create very compact RTL. Example 11-6 shows the C code for one of two key nested loops for the DCT algorithm:

```
/* Horizontal */
for(y=0;y<8;y++){
  for(x=0;x<8;x++)
    reg[x]=0;

  for(x=0;x<8;x++){
    for(u=0;u<8;u++){
      v=block[y][x]*c[x][u];
      v+=2048;
      v>>=12;
      if(s[x][u]) v=-v;
      reg[u]+=v;
    }
  }
}
```

Example 11-6

Example 11-7 is the RTL for this loop, coded as a state machine using *istates*.

```
//----------- horizontal processing --------------
sub_state_machine horizontal_dct();
  int v;
  begin
    done = 0;
    for (istate y=0;y<8;y++) begin
      for(istate x=0;x<8;x++) begin
        for(int u=0;u<8;u++) begin
          v=pixin*c[x][u];
          v+=2048;
          v>>>12;
          if(s[x][u]) v=-v;
          if (x ==0) reg_x[u]<=v;
          else if (x < 7) reg_x[u]<=reg_x[u]+v;
          else tmp_x[u] <= reg_x[u] + v;
        end
      end
    end
    done = 1;
  end
endstate_machine
```

Example 11-7

Note that:

1. The three main loops (*y, x, u*) are virtually identical to the C code. Only a minimum amount of changes are needed to convert the high level C to RTL.
2. Using *for (int u...)* in the inner loop specifies that all eight *u* values for a given pixel will be calculated at the same time – that is, in the same clock cycle by parallel hardware.
3. Using *for (istate y...)* and *for (istate x ...)* specifies that each pixel in the 8x8 input array will take one state (one cycle in this case) to process.
4. The key modifications that make this version more RTL-like than the C code are the lines where *reg_x* and *tmp_x* are assigned values. This form of the code achieves minimum latency – and avoids having to initialize *reg_x* to 0 (as is done in the C code).

This use of the for-istate capability starts to blur the line between high level synthesis and RTL. With minor modifications, we can convert the C code to RTL. The key modifications consist of specifying which loops get unrolled (for-int loops) and which loops don't get unrolled (for-istate loops). In extended SystemVerilog, we indicate this in the source code; in C-synthesis this is typically specified in pragmas or in the GUI.

Fork and Join

As part of the state machine extensions to SystemVerilog, we propose making the SystemVerilog construct *fork-join_none* synthesizable. This construct makes it possible to specify pipelines explicitly and concisely. In SystemVerilog, *fork* causes processes to be spawned in parallel; *join_none* specifies that the forking process does not wait for the forked process to complete before continuing execution.

In Example 11-8, we outline how this can be done for a DCT. In this example, each state machine calls the next state machine in the pipeline, at the appropriate time.

The *horizontal_dct* state machine processes eight pixels, then forks the state machine *mem_write*. *Horizontal_dct* continues processing the next eight pixels (one pixel per clock cycle) while *mem_write* writes the eight processed pixels into memory, one pixel per clock cycle.

When enough pixels have been written into the memory, *mem_write* forks off *mem_read*.

When *mem_read* has read the first pixel, it forks the *vertical_dct* state machine.

Once the *vertical_dct* has completed processing one pixel, it forks *output_pixels*.

At that point, all five processes (*horizontal, mem_write, mem_read, vertical, output_pixels*) are operating in parallel, as a pipeline. The start-up of the pipeline is shown in Figure 11-1.

The classic problem of pipeline stalls is handled by how we interpret the *fork* command. For instance, the *horizontal_dct* forks *mem_write* every time through its *y*-loop; that is, each time it is done with a set of eight pixels. If the previous iteration's *mem_write* has not completed – that is, if it is not done writing the eight previous pixels to memory – then *horizontal_dct* stalls, waiting for *mem_write* to complete. That is, the *fork* command is blocking – if the forked process is not done, the forking process waits until it is done.

```
//------------ horizontal processing -----------
sub_state_machine horizontal_dct();
  done = 0;
  while (run) begin
    for (istate y=0;y<8;y++) begin
      for(istate x=0;x<8;x++) begin
        for(int u=0;u<8;u++) begin
          v=pixin*c[x][u];

            ...

        end
      end
      fork mem_write(y); join_none // new feature
    end
  end
  done = 1;
endstate_machine

//------------- write scratchpad memory ---------
sub_state_machine mem_write(int y);
  for(istate x=0;x<8;x++) begin
  ...
  if ((y == 7) && (x == 0) fork mem_read(); join_none
end
endstate_machine

//---------- read scratchpad memory ------------
sub_state_machine mem_read();
  for (istate y=0;y<8;y++) begin
    for(istate x=0;x<8;x++) begin

      ...

      if ((y == 0) && (x == 0))
    fork vertical_dct; join_none
    end
  end
endstate_machine

//------------ vertical processing -------------
sub_state_machine vertical_dct();
int v;
  for(istate y=0;y<8;y++) begin
```

Example 11-8

```
   for(istate x=0;x<8;x++) begin
     for(int u=0;u<8;u++) begin
       v=v_pixin*c[x][u];
             ...
     end
     fork output_pixels(); join_none
   end
 end
endstate_machine

//------------- output results ---------------
sub_state_machine output_pixels();

   ...
 endstate_machine
```

Example 11-8 (continued)

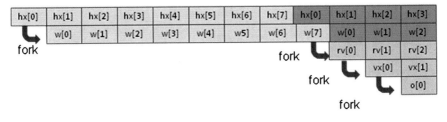

Figure 11-1 The DCT pipeline generated by *fork-join_none*.

The combination of *for-istate* and *fork-join_none* is a major step in closing the gap between RTL and high level synthesis. We are still in the RTL domain, since the compiler does not do any scheduling, and does not move operations from one clock cycle to another. The cycle-by-cycle behavior of the design is completely determined by the source code. However, the concise, C-like style of coding is much higher in abstraction than ordinary RTL.

Summary

By extending and improving the synthesizable subset of SystemVerilog, we effectively raise the level of abstraction of RTL design, while remaining entirely within the RTL paradigm. By adding some new primitives to SystemVerilog, we create a more natural and concise way of describing the most common RTL structures. In doing so, we make RTL code:

1. Easier to get right the first time
2. Easier for engineers to review and analyze (in design reviews, etc.)
3. Easier to develop automated tools to analyze and detect errors
4. Easier to verify

In particular, the proposed constructs make RTL code easier to understand and analyze – both for humans and for EDA tools - because they:

1. Eliminate inferencing of registers – flip-flops and latches are explicitly declared
2. Eliminate inferencing of clocks and reset – they are explicitly declared
3. Eliminate inferencing of state machines – they are explicitly declared, allowing improved optimization (during synthesis) and improved verification through state coverage analysis
4. Eliminate a class of synthesis-simulation mismatch problems by removing the confusing simulation behavior of blocking vs. non-blocking assignments

There is No Substitute for Good Code

The extensions outlined above are the first steps in making the synthesizable subset of SystemVerilog into a genuine domain-specific language for hardware design. These are only first steps; other constructs, including FIFOs and packets, are needed.

But improving the language is only part of the solution to improving design productivity. Ultimately, it is how we use the language that determines the quality and productivity of design. There is no substitute for good code. Said differently, any language can be abused to produce unreadable, unsupportable, and unverifiable designs.

The key to good code has always been structure. In 1968. Edsger W. Dijkstra published the famous paper "Go To Statement Considered Harmful"[14], and a revolution in programming was born. Based on the work of Dijkstra and other researchers, programmers moved from what today is considered "spaghetti code" to structured programming.

In the last twenty years or so, object oriented programming has extended the concepts of structured programming, to help deal with really large programs. With object oriented programs, functionality is partitioned into separate objects that provide services but hide the internal implementation.

In RTL design, we use modules to provide the partitioning and information hiding provided in software by classes. A module with a well-designed interface hides its internal structure from the rest of the system, and provides services (typically some form of data processing) via its interface. A poorly designed interface exposes the internal structure of the module to the rest of the system, undermining the structure of the design and compromising overall quality, robustness, and supportability.

But if we look inside the module, at the RTL code itself, we typically see code that seems to violate all of the recommendations of structured programming. Most

RTL is a random collection of combinational and sequential processes, without any interconnecting structure. The code consists of dozens or even hundreds of independent small objects (processes) that are connected by signals, but whose interconnections are virtually impossible to understand without drawing some kind of diagram or schematic.

The methodology developed by Dijkstra included the practice of separating a program into separate sections, each of which has a single point of access. In C, "main" provides this single point of access – once we understand what main does, and what is done by the functions called by main, we understand the program.

But typical RTL does not have a main. All the processes are on equal footing. To understand the program, we must read and memorize the entire program and then sort out the relationships of all the processes – how they work together to implement a function.

The coding style described in this book attempts to put RTL code squarely within the category of structured code. The features described in this chapter can make this significantly easier. But the RTL designer must understand and practice the basic principles of structured code:

- Each module must have a single entry point that allows a systematic way to review of the code.
- Functions and tasks can be used to encapsulate complex code; they are superior to processes because they are called. The call graph of the module provides a clear description of the structure of the module.

But most importantly of all, the RTL designer must keep the complexity of the code well within the intellectual capacity of the designer, various reviewers, and any designer who may have to support the code in the future.

Chapter 12
The Future of Design

> The Fundamental Principle of Structured Programming is that
> at all times and under all circumstances, the programmer must
> keep the program within his intellectual grasp..
>
> —Tom Harbron

In this final chapter, we move away from the discussion of particular design techniques to a discussion of design in general, and future trends in SoC design.

The central problem in designing electronic products is the problem of scaling. As technology relentlessly executes Moore's law, the design techniques and practices of one generation quickly become obsolete. They are no longer adequate to deal with the complexity of the next generation of chips, software, systems, and end products.

To anticipate what chip design will be like in the future, we need to look at how the core technologies are scaling.

The semiconductor industry has been impressive in its ability to modify processes, materials, and equipment to keep shrinking technology nodes. Copper interconnect and hi-k dielectric/metal gates are just two of many examples of innovations that have kept the industry tracking Moore's law. There are significant challenges in scaling CMOS below 20nm, but techniques such as FinFETs, Extreme Ultraviolet lithography, and direct e-beam write are promising approaches to meeting these challenges.

The EDA industry also has a good track record of meeting the challenges of implementing complex chips. Today's place and route tools are much more capable than a decade ago. They have taken advantage of faster processors and are using parallel programming techniques to take advantage of multi-core processors. The introduction of many resolution enhancement techniques, including Optical Proximity Correction (OPC), Sub Resolution Assist Features (SRAF) and support for Phase Shift Masks (PSM) have allowed optical lithography to continue scaling deep into the nanometer range.

M. Keating, *The Simple Art of SoC Design: Closing the Gap between RTL and ESL*, DOI 10.1007/978-1-4419-8586-6_12, © Synopsys, Inc. 2011

Design

Unfortunately, design – coding, verifying, and synthesizing functionality – is not doing nearly as good a job of scaling as semiconductor process and implementation tools. The basic tools and methodology for RTL-based design have changed little over the last fifteen years. There have been small improvements in the synthesizable subset of RTL, as Verilog has matured from Verilog95 to Verilog 2k to SystemVerilog. But the actual code as written by most engineers has changed little.

Similarly, verification-specific languages, like the verification side of SystemVerilog, have facilitated writing larger and more complex test benches. But the fundamental approach to verification – directed tests plus constrained random testing of RTL – has remained unchanged for at least ten years.

Synthesis, as well, has changed little over the last 10 years. C-based synthesis has been around, and had some limited success, over that whole time. But it has not become anything close to a mainstream technology. And it is not likely to do so as long as it remains in its current form. RTL synthesis has improved in runtime and capacity, but little of its fundamental capabilities have changed.

The result is that teams have scaled the design and verification process by growing the size of the team as chips become larger and more complex.

Function Does Not Scale

Figure 12-1 from the International Technology Roadmap for Semiconductors shows how hardware and software design productivity has lagged Moore's Law.

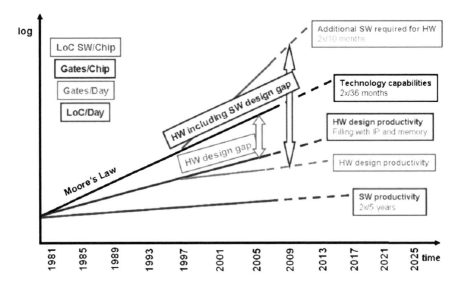

Figure 12-1 Semiconductor Industry Association. The International Technology Roadmap for Semiconductors, 2009 Edition. International SEMATECH:Austin, TX, 2009. Used by permission.

A great example of the design gap is this:

- Since the 1980's, semiconductor technology has gone from resolving 10 micron features to resolving 22 nm features (a factor of about 500). They can now deposit a layer of material precisely one atom thick on a substrate.
- Since the 1980's, software productivity has gone from about 30 lines of code a day (per engineer, over the life of a project) to about … 30 lines of code a day.

We live in a world where function (hardware and software) is described in code. But code does not scale. Individual coders cannot code more lines of code (RTL or C++) than they could decades ago. And as projects get bigger, productivity actually decreases. One engineer can code about 10,000 lines of (debugged, production-ready) code in a year. But 100 engineers cannot code 1,000,000 lines of (debugged, production-ready) code in a year – the problems of coordinating work and debugging complex problems degrade productivity[15][5].

The functionality of the 30 lines of code (per engineer-day) has increased very slowly over time, and most of this increase has been due to the use of libraries and IP. The introduction of C++ in 1983 also gave software productivity a small, one time gain.

There is no evidence that RTL coders have done any better than their software colleagues. RTL productivity has been roughly constant at (on the order of) 100 gates/30 lines of code per engineer-day (over the life of the project, including verification). There is some indication that this productivity has decreased as IP designs become bigger and more configurable – hence harder to design and verify.

Another telling point about code-based design is quality. Studies ([4a][4b][4c]) have shown that coders inject about one bug for every ten lines of code. That means for a 10,000 line program (or piece of RTL code), about 1,000 bugs must be removed through testing/verification.

We don't have good data on RTL designs, but for software we know [5] that the very best software teams ship code with about 1 defect per thousand lines of code. That is, they find and remove about 99% of the bugs in the original code. This is a remarkable accomplishment, but it still leaves one bug per thousand lines of code: a minor problem for a 10,000 line program, but a major headache for a program of 1 million or 10 million lines of code.

So code does not scale: productivity and quality have remained roughly constant for at least twenty years, at 30 lines of code per engineer-day, 1 bug per 10 lines of code initial quality, 1 bug per 1,000 lines of code shipped quality.

Now let's consider how the size of designs has changed over the last few years. From the IEEE Spectrum:

By Robert N. Charette // February 2009

The F-35 Joint Strike Fighter, scheduled to become operational in 2010, will require about 5.7 million lines of code to operate its onboard systems. And Boeing's new 787 Dreamliner, scheduled to be delivered to customers in 2010, requires about 6.5 million lines of software code to operate its avionics and onboard support systems.

These are impressive amounts of software, yet if you bought a premium-class automobile recently, "it probably contains close to 100 million lines of software code," says Manfred Broy, a professor of informatics at Technical University, Munich, and a leading expert on software in cars.

Figure 12-2 100 Million Lines of Code in a High End Car (Photo courtesy of John Filiss, www.seriouswheels.com).

To give an example from IP design: The Verification IP (VIP) for the USB is the bus functional model that drives transactions into the RTL model for the USB. It is a critical part of the overall test bench for the USB digital IP. For USB2.0 this VIP was 141,000 lines of code. For USB3.0 it is 272,000 lines of code. That is, from one generation to the next, it has grown by about 2x, to over a quarter of a million lines of code. All this is just to create USB transactions.

The total code for the USB3.0 – RTL and test bench – is about 900,000 lines of code.

Just for reference, the Encyclopaedia Britannica has about 4 million lines (44 million words)

Small is Beautiful – and Tractable

The reality is this: we are very good at small designs, and very bad at large ones.

In particular, we are very good at analyzing and solving problems when we can see the entire problem at the same time. For instance, if we were planning a trip and looked it up on a map, we would want the map (at least initially) to show the entire route. Tracking a route that covers three or four pages of a map is much, much harder. After planning the overall route, we probably want a detailed map of the destination – how to get from the freeway to a particular hotel, for example. But this is a secondary consideration, done only after the overall route has been planned.

Top down design, of course, is exactly this practice. The top level algorithm is described in a few lines, relying on subsidiary functions. In software classes, we are taught that a function should fit on a single screen.

The reason for this is that, smart as humans are, we still have a very limited ability to reason about complex problems. We reason well about what we can see. We reason poorly about what we can't see – but instead have to remember.

Unfortunately, the one-screen rule (the most basic programming rule in theory) is the one most often violated in practice. Any solution to the challenges of design productivity and quality must address this issue. As long as we are dealing with units of code that are too large to understand, we are in trouble.

How Does IP Help?

The reuse revolution has certainly helped design productivity at the chip level. IP such as processors, memory, memory controllers, USB and PCI interfaces have made chip design largely a question of assembling IP and designing the interconnect and communication between IPs. Subsystems consisting of multiple IPs – such as video subsystems – enable a natural and effective hierarchy for chip design.

But if we look below the level of the subsystem, and look at the IP itself, we see a problem. Table 12-1 shows the basic hierarchy of digital design, from the lowest level on up.

Table 12-1 Abstraction levels in SoC design.

Type	Example	Size
Standard Cell	Scan Flop	10's of gates
Small IP	FIFO Controller	400 lines of code
Large IP	USB 3.0	182,000 lines of code
Subsystem	Video Subsystem	10's of IPs
SoC	Cell Phone Chip	10's of subsystems

The obvious discontinuity here is the jump in size from standard cells to small IP to large IP. With standard cells and with subsystems and chips we are assembling larger blocks out of a reasonable number (~10) of smaller blocks. This is a rational form of composition.

But a large IP that consists of 182,000 lines of code is a huge jump from the size of a standard cell or small IP. It is the power and verbosity of RTL that gets us into this kind of untenable scaling.

For example, another large IP – a PCI Express controller – has a total of 53,000 lines of code. The top level file consists of 1500 lines of code distributed as:

- Input and output declarations: 120 lines
- All declarations (including input and output): 600 lines
- Assign statements: 80 lines
- Instantiations: 900 lines
- Number of instantiations: 9

So the top level of the PCI Express takes 1500 lines to instantiate just 9 modules.

Automation and Scaling

In general, the problem of scaling has been most effectively addressed by automation: that is, turning specific designs problems into generic problems that can then be solved in a routine, automated fashion.

For example, when chips consisted of a few thousand gates, it was rational to design each transistor by hand, and thus to optimize performance and area. But as chips grew to tens and then hundreds of thousands of transistors, this kind of custom design became economically infeasible for most designs. Designers then turned the specific problem of designing an optimal transistor into the generic problem of developing a standard cell library.

Semiconductor companies have always taken this approach as well – developing complex processes that use "step and repeat" to allow the manufacture of very complex devices in high volume.

EDA has similarly turned place and route – once a manual process – into a fully automated process, using sophisticated algorithms to turn the specific problem – how to route this chip – into the generic problem of solving an optimization problem.

In all these cases, engineers developed infrastructure that enabled automation. This infrastructure can be very expensive – fabs can cost billions of dollars, EDA tools are expensive to develop and to purchase. But the payoff has been very high.

Engineers have not made such infrastructure investments for RTL design (or, with very few exceptions, for software design). Instead, most engineering organizations have adopted a "code like hell" approach to development. The result has been an explosion in code size and in the cost of chip development.

The only way to reduce development cost and allow RTL design to scale is to invest in infrastructure. We need technology that will significantly reduce the amount of code required to describe the function of a digital design. And we need technology that enables engineers to visualize complex design, not by scrolling through pages and pages of code, but by examining a single, static representation of the design.

The Future of Design

The future of design will be determined by the ability of designers and EDA companies to develop and use scalable design techniques.

Data structures will need to become more powerful, and timing control will have to rise in abstraction. More and more timing will be left out of the code, especially for algorithm-intensive code. The timing that remains explicit in the code will be controlled by hierarchical state machines that allow the designer to express concurrency and pipelining naturally and simply.

High level system modeling (typically in SystemC) has long used the concept of Transaction Level Modeling – where system level communication is modeled by passing data structures (or in TLM 2.0, pointers to data structures) from one block (IP or subsystem) to another.

As interfaces become more powerful and more standardized (data structures passed by FIFOs locally and busses globally), the RTL interface and the TLM interface will start to merge as a concept. Designers will be able to code interfaces as data structures and guide the compiler to interpret them either as TLM objects (for high level modeling) or as RTL interfaces (for synthesis).

All of this, of course, means that design will be rising in abstraction. Designers will write code that more naturally describes their intent, and compilers will play an extended role in scheduling and optimization.

Verification

Raising the abstraction of design will have a significant impact on verification. The complexity of today's applications, and the fallibility of human engineers, will mean that complete verification will always be an elusive goal. But by raising the level of abstraction, we can make the verification process simpler, more robust, and more automated.

There is a limit, of course, to how high we can raise the abstraction of a design. The code must reflect the intended behavior of the design. At the highest level of abstraction, the code is really just an executable specification. This executable specification cannot be avoided: at some point, engineers must take the marketing requirements, and various other natural language specifications, and turn them into something that can be simulated and tested. This translation of requirements into code is an activity that can't be avoided, and which has to be done by humans.

Today, the code we write is typically much lower level than required for an executable specification. Typically it has much more detailed timing information, to meet the requirements of the synthesis engine. But if we improve the synthesis tools to the point where the executable specification can be synthesized, then we change the nature of verification.

Verification of the execution specification, like writing it, is an unavoidable, fundamentally human activity. We may make mistakes, but the mistakes are in our understanding of what the design is expected to do, not in the "incidental" aspects of the design such as low level coding decisions.

Once we have a verified executable specification, we should then be able to automate the detailed design, through high level synthesis. Verification can then be automated: formal verification can prove that the detailed implementation is equivalent to the executable specification. At that point, verification is as automated and complete as possible.

Equivalence checking between high level (untimed) code and RTL is a difficult problem, but enough progress has been made recently [16a][16b][16c] to suggest it could become a mature technology in a few years.

Visualization

One key element of managing code complexity is to reduce code size. All the techniques discussed in this book focus on this approach. But improved visualization tools can also help manage complexity.

Visual design tools – like UML and VisualHDL – have met with limited success at best. Designers really do prefer to enter code rather than make drawings. But visualization tools such as waveform viewers are essential tools for debug. IDEs like Eclipse and VisualStudio are very popular because they help software engineers understand their code. The state machine visualization capability of Debussy is also very useful.

There is a great opportunity to extend these tools and enable designers to see (on a single screen) an entire design – at least at some level of abstraction. The tools that would be useful include:

- A state machine display tool that takes advantage of the proposed state machine construct to display hierarchical state machines in a useful way. We could imagine clicking on a state at the top level and popping up a display of the sub-state machine. It could also measure the complexity of the state machine.
- A connectivity diagram tool that would show the connections and interfaces between modules. With wires this becomes impossibly messy; but with structs and FIFOs as the interconnection mechanism, we can envision a very useful display. It could also measure and report the complexity of the overall design.

Drivers of the Solution

As noted earlier, code has not scaled over the last ten or fifteen years. We have not been able to keep the number of lines of code (of design and testbenches) close to constant as complexity increased. Instead, the number of lines of code per gate or per function has remained roughly constant, driving design cost up as gate count has increased.

The proposals in this book suggest a path for managing complexity, reducing the number of lines of code, and getting code to start scaling. The question is: what is going to drive the changes required? These changes will require investments from EDA companies to develop and improve tools, and from designers as they migrate from current practices to new and unfamiliar ones.

There are a number of factors working against these investments. The two most recent revolutions in design – synthesis and IP reuse – were driven out of financial opportunity. Designers saw that they could reduce their development schedule and costs by adopting them, and were willing to pay for this advantage. The result was a significant increase in revenue to the companies that provided synthesis tools and IP.

But the world has changed over the last decade or so. Growth in the semiconductor industry has slowed dramatically, and it is maturing as an industry. That means that there is more focus on reducing costs, and in particular managing tool budgets carefully. The result is that EDA revenues are staying roughly constant, with little opportunity for significant growth. So it is hard for the large EDA companies to make major investments in new technology. There is just not enough upside reward to justify the risk.

The alternative path of having small EDA startups develop innovative technology gets more difficult each year. With the recent downturn, funding has been difficult to find. More importantly, it is hard to produce a compelling EDA product in isolation from the other tools and flows used in design. The cost of adopting a new tool for the designer is very high if the tool is not already integrated into the design flow. The result is that innovative EDA startups need considerable investment capital (which is hard to find) to develop products that are difficult for the customers to adopt (and hence do not garner high prices).

Thus, in the short term, market forces are not going to drive major investments in innovation. But there are several potential crises that could create a compelling need to solve these problems.

The first kind of crisis would be for major semiconductor companies to find that they simply cannot make the products that their customers demand. If the complexity of design becomes such that chips cannot be designed to an acceptable level of quality in an acceptable amount of time for an acceptable price, then they may demand rapid improvements of how code is developed and verified. This crisis is slowly building now, but it is a "frog in the pot" kind of crisis. (There is an old saying that if you toss a frog in a boiling pot, it will jump out. But if you toss him in a cold pot and then slowly increase the temperature, it will never jump out – just end up cooking.) The productivity challenge is turning up the temperature on the semiconductor industry, but it is showing no signs of jumping yet.

The other kind of crisis is if code quality ends up killing people. If a major car company were to find that its embedded software had bugs that were leading to fatal crashes, and that it could not fix these bugs reliably, then this too would precipitate change.

This second kind of crisis has to do with the fact that more and more chips and software are finding their way into mission-critical applications. Traditionally, mission-critical applications have used a different methodology from mainstream applications. Considerably more testing and verification, more design discipline and design rules, and more careful and thorough review have been used for flight systems, for example, and space systems. This methodology achieves a much higher level of quality than mainstream practices, but at a considerable cost. As mentioned in Chapter 1, NASA can achieve a remarkable quality level of about .004 defects per KLOC, but at a cost of nearly \$1,000 per line of code (compared to \$25 per line of code with mainstream methodologies.)

But now cars, medical devices, and other complex systems are incorporating chips and software. If these systems are developed using commercial (as opposed to mission-critical) development processes, then there is a significant chance of bugs creating serious problems.

On a more optimistic note, individual engineers in many companies understand and are working on these problems. The technical community, both in hardware design and in software design, is actively discussing how to move our code-based technology forward. Hopefully, when the financial imperative occurs, we will have the fundamental approaches for a solution worked out, and will be able to quickly implement the solutions.

Summary

We live in a world where function is described in code, and that is not likely to change.

Code size counts – after a certain size, code becomes intractably hard to understand and debug. As design complexity constantly increases, we must constantly be developing techniques for reducing code size.

This book has attempted to show how to reduce code size – and complexity – in the small but important domain of synchronous digital design.

Appendix A
Guidelines

This chapter summarizes the design and coding guidelines from previous chapters. It also provides some guidelines for coding datapaths using signed data types.

General

Guideline: Designs and coding should be kept as simple as possible, consistent with the objectives of the design. The best measure of simplicity/complexity is the size of the state space of the design.

Guideline: Locality. Related information should be located together in the code.

Guideline: File Size. Source code file should be no more than 5 pages (about 300 lines).

Guideline: Rule of Seven. Any design or part of a design should consist of at most seven to nine objects.

State Machines

Guideline: There should be at most one state machine per module.

Guideline: All sequential code (registers that do any computation) should be in a state machine. Registers that simply buffer a signal do not need to be in a state machine.

Guideline: A state machine should be coded as a sequential process (**always_ff**). Auxiliary **always_comb** combinational processes can be used to drive combinational outputs if required.

M. Keating, *The Simple Art of SoC Design: Closing the Gap between RTL and ESL*, DOI 10.1007/978-1-4419-8586-6, © Synopsys, Inc. 2011

Guideline: A state machine with more than 7 to 9 states should be coded hierarchically. Any single sub-state-machine should have at most 7to 9 states.

Processes

Guideline: Use the SystemVerilog constructs **always_ff** and **always_comb** rather than **always@**. These constructs remove the ambiguity inherent in **always@**, where we can accidently code a latch when we intend a flip-flop.

Combinational Code

Guideline: Combinational code should be coded as functions.

Guideline: Functions in RTL should be coded as **automatic** to avoid synthesis/simulation mismatches.

Guideline: Where practical, the function should be written with arguments that include all the signals needed by the function. But if the number of arguments gets large, this makes the code much harder to read; in this case, the function can be written without arguments and the function code can refer to global signals directly.

Hint: Debugging with automatic functions: When using a waveform viewer to debug code written using functions, we need to be a bit careful. Normal combinational code (**always_comb** blocks) are updated on the waveform viewer whenever an input changes. Functions are updated only when they are called. But with a little experience, debugging with functions becomes quite straight-forward.

Data Structures

Guideline: Use **struct**s to assemble signals into data structures.

Guideline: Use enumerated types.

Guideline: Avoid **reg** and **wire**.

Guideline: Use **bit** and **logic** only as components of a **struct**.

Interfaces

Guideline: Design interfaces that isolate the two modules they connect. The FIFO is a great example of an interface that isolates the timing between two modules.

Guideline: Use the **interface** or **struct** constructs rather than individual signals. (See Appendix D for a discussion of the **interface** construct.)

Guideline: Limit the number of interfaces of a module to 7-9.

Guideline: Design the interface to consist of a command word and a data word, rather than random wires.

Guideline: Use enumerated types to define the valid values of the command word.

Coding

Guideline: use "**if** (a)" instead of "**if** (a == 1)"

Partitioning

Guideline: Any module that has significant computation should be in a module that uses a single clock and a single reset. This facilitates analysis, since only one clock domain needs to be analyzed at a time.

Guideline: Any module with multiple clocks should consist only of the logic necessary for crossing the two clock domains. This facilitates analysis of the clock domain crossing logic. This logic can be quite tricky, so isolating it in a very small module can make analysis much easier.

Guideline: Modules that contain logic (and synchronizers) for crossing clock domains should be named with a prefix such as SYNC to indicate the special role of the module. Using such a prefix helps implementation engineers identify critical clock-crossing paths in synthesis reports and dynamic netlist simulations.

Guidelines for Datapath Synthesis

The following guidelines are excerpted from a set of datapath guidelines; the full set can be found online [28].

These guidelines were originally developed to guide designers on how to get the best quality of results (QOR) from synthesis. But what they really amount to is this: to get the best results, do NOT try to out-guess the synthesis tool by doing manual pre-synthesis (like converting a multiply to a shift and add). Write the code in the most general fashion (that is, leave it as a multiply) and let the synthesis tool do the optimizations.

This general approach fits in with the observations made earlier in the book. Considering the complexity of design, and the strength and maturity of the synthesis tools, we need to code for readability by human beings. The tools are smart enough to convert this user-friendly code into an optimal gate netlist.

The guidelines below have been re-written slightly from the versions on the solvenet website, in order to emphasize how they improve the readability of code and to comply with the other guidelines in this book.

Note: to keep the following examples simple, we use *assign* statements. In real code, we would expect the datapath arithmetic to be larger and more complex and to be written in a *function*.

Note: A lot of these recommendations concern signed arithmetic. Signed arithmetic was added in Verilog2001 and can be a bit tricky. Hopefully the guidelines below will help designers to avoid problems.

Signed Arithmetic

- **Rule:** Use type 'signed' (Verilog 2001, SystemVerilog) for *signed/2's complement arithmetic* (that is, do not emulate signed arithmetic using unsigned operands/operations). Also, do not use the 'integer' type except for constant values.
- **Rationale:** Simpler, cleaner code and provides better QOR.
- **Example:** Signed multiplication.

Not Recommended	Recommended
| ```
input    [7:0] a, b;
output [15:0] z;

// a, b sign-extended to
width of z

assign z = {{8{a[7]}}, a[7:0]} *
           {{8{b[7]}}, b[7:0]};

// -> unsigned 16x16= 16 bit multiply
``` | ```
input  signed  [7:0] a, b;
output signed [15:0] z;




assign z = a * b;



// -> signed 8x8=16 bit multiply
``` |
| ```
input    [7:0] a, b;
output [15:0] z;

// emulate signed a, b

assign z = (a[6:0] - (a[7]<<7)) *
           (b[6:0] - (b[7]<<7));

// -> two subtract +
unsigned //16x16=16 bit multiply
``` | ```
input              [7:0] a, b;
output             [15:0] z;
wire    signed [15:0] z_sgn;



assign z_sgn = $signed(a) *
                $signed(b);
assign z     = $unsigned (z_sgn);

// -> signed 8x8= 16 bit multiply
``` |

Sign-/Zero-extension

- **Rule:** Do not manually *sign-/zero-extend* operands if possible. By using the appropriate unsigned/signed types, correct extension is done automatically.

- **Rationale:** Simpler, cleaner code and better QoR because synthesis can more easily/reliably detect extended operands for optimal implementation.
- **Example:**

Recommended
| ```
input  signed [7:0] a, b;
output signed [8:0] z;
```<br><br>*// a, b implicitly sign-extended*<br><br>```
assign z = a + b;
``` |

Mixed Unsigned/Signed Expression

- **Rule:** Do not mix *unsigned and signed* types in one expression.
- **Rationale:** *Unexpected behavior / functional incorrectness* because Verilog interprets the entire expression as unsigned if one operand is unsigned.
- **Example:** Multiplication of unsigned operand with signed operand

Functionally incorrect	Functionally correct
| ```
input         [7:0] a;
input  signed [7:0] b;
output signed [15:0] z;
```<br><br>*// expression becomes unsigned*<br><br>```
assign z = a * b;
```<br><br>*// -> unsigned multiply* | ```
input         [7:0] a;
input  signed [7:0] b;
output signed [15:0] z;
```<br><br>*// zero-extended, cast to signed (add '0'*<br>*// as sign bit)*<br>```
assign z = $signed
            ({1'b0, a}) * b;
```<br><br>*// -> signed multiply* |
| ```
input  signed [7:0] a;
output signed [11:0] z;
```<br><br>*// constant is unsigned*<br><br><br>```
assign z = a * 4'b1011;
```<br><br>*// -> unsigned multiply* | ```
input  signed [7:0] a;
output signed [15:0] z1, z2;
```<br><br>*// cast constant into signed*<br>```
assign z1 = a * $signed
             (4'b1011);
```<br><br>*// mark constant as signed*<br>```
assign z2 = a * 4'sb1011;
```<br>*// -> signed multiply* |
```

Signed part-select / concatenation

- **Note:** *Part-select* results are unsigned, regardless of the operands. Therefore, part-selects of signed vectors (for example, "a[6:0]" of "input signed [7:0] a") become unsigned, even if part-select specifies the entire vector (for example, "a[7:0]" of "input signed [7:0] a").
- **Rule:** Do not use part-selects that specify the entire vector.
- **Note:** *Concatenation* results are unsigned, regardless of the operands.
- **Example:**

Functionally incorrect	Functionally correct

`input  signed  [7:0] a, b;`
`output signed [15:0] z1, z2;`
 	
// a[7:0] is unsigned -> zero-extended
 	
`assign z1 = a[7:0];`
 	
// a[6:0] is unsigned -> unsigned multiply
 	
`assign z2 = a[6:0] * b;`
	`input  signed  [7:0] a, b;`
	`output signed [15:0] z1, z2;`
	
	// a is signed -> sign-extended
	
	`assign z1 = a;`
	
	// cast a[6:0] to signed -> signed
	// multiply
	
	`assign z2 = $signed(a[6:0]) * b;`

Expression Widths

- **Note:** The width of an expression in Verilog is determined as follows:
  - ° *Context-determined expression*: In an assignment, the left-hand side provides the context that determines the width of the right-hand side expression (that is, the expression has the width of the vector it is assigned to).
    Example:

```
input  [7:0] a, b;
output [8:0] z;

assign z = a + b;   // expression width is 9 bits
```
```
input  [3:0] a;
input  [7:0] b;
output [9:0] z;

assign z = a * b;   // expression width is 10 bits
```

° *Self-determined expression*: Expressions without context (for example, expressions in parentheses) determine their width from the operand widths. For arithmetic operations, the width of a self-determined expression is the width of the widest operand.

Example:

Unintended behavior	**Intended behavior**
| ```
input   signed   [3:0] a;
input   signed   [7:0] b;
output           [11:0] z;
```<br><br>*// product width is **8** bits (not 12!)*<br><br>```
assign z = $unsigned(a * b);
```<br><br>*// -> 4x8=**8** bit multiply* | ```
input   signed   [3:0] a;
input   signed   [7:0] b;
output           [11:0] z;
wire    signed [11:0] z_sgn;
```<br><br>*// product width is **12** bits*<br><br>```
assign z_sgn = a * b;
assign z     = $unsigned(z_sgn);
```<br>*// -> 4x8=**12** bit multiply* |
| ```
input    [7:0] a, b, c, d;
output         z;
```<br><br><br><br>```
assign z = (a + b) > (c * d);
```<br><br>*// -> 8+8=**8** bit add + 8x8=**8** bit*<br>*// multiply + **8**>**8**=1 bit compare* | ```
input    [7:0] a, b, c, d;
output         z;
wire     [8:0] s;
wire     [15:0] p;
```<br><br>```
assign s = a + b;// -> 8+8=9 bit add
```<br><br>```
assign p = c * d; // -> 8x8=16 bit
                  //   multiply
```<br><br>```
assign z = s > p;  // -> 9>16=1 bit
                   //   compare
``` |
| ```
input   [15:0] a, b;
output  [31:0] z;
```<br><br>```
assign z = {a[15:8] *
b[15:8], a[ 7:0] * b[ 7:0]};
```<br><br>*// -> two 8x8=**8** bit multiplies,*<br>*// bits z[31:16] are 0* | ```
input   [15:0] a, b;
output  [31:0] z;
wire    [15:0] zh, zl;
```<br><br>```
assign zh = a[15:8] * b[15:8];
assign zl = a[ 7:0] * b[ 7:0];
assign z  = {zh, zl};
```<br><br>*// -> two 8x8=**16** bit multiplies* |

° *Special cases*: Some expressions are not self-determined even though they seem to be. Then the expression takes the width of the higher-level context (for example, the left-hand side of an assignment).

Example: Concatenation expression

Cluster Datapath Portions

• **Rule:** *Cluster related datapath portions* in the RTL code into a single combinational block. Do not separate them into different blocks. In particular,

- ° Keep related datapath portions within *one single hierarchical component*. Do not distribute them into different levels or subcomponents of your design hierarchy.
- ° Do *not place registers* between related datapath portions. If registers are required inside a datapath block to meet QoR requirements, use retiming to move the registers to the optimal requirements, use retiming to move the registers to the optimal location *after* the entire datapath block has been implemented (see [28] for more details).

- **Rationale:** Simpler, cleaner code and better QoR because bigger datapath blocks can be extracted and synthesized.

Component Instantiation

- **Rule:** Do not *instantiate arithmetic DesignWare components* if possible (for example, for explicitly forcing carry-save format on intermediate results). Write arithmetic expressions in RTL instead.
- **Rationale:** Simpler, higher level code. Better QoR can be obtained by exploiting the full potential of datapath extraction and synthesis.
- **Example:** Multiply-accumulate unit

Bad QoR	Good QoR

``
input   [7:0] a, b;
input   [15:0] c0, c1;
output [15:0] z0, z1;
wire   [17:0] p0, p1;
wire   [15:0] s00, s01, s10, s11;
 	
// shared multiply with explicit carry-
 	
// save output
DW02_multp #(8, 8, 18) mult (
  .a(a), .b(b), .tc(1'b0),
  .out0(p0), .out1(p1));
 	
// add with explicit carry-save output
DW01_csa #(16) csa0 (
  .a(p0[15:0]), .b(p1[15:0]), .c(c0),
  .ci(1'b0), .sum(s00), .carry(s01));
DW01_csa #(16) csa1 (
  .a(p0[15:0]), .b(p1[15:0]), .c(c1),
  .ci(1'b0), .sum(s10), .carry(s11));
 	
// carry-save to binary conversion
(final adder)
DW01_add #(16) add0 (
  .A(s00), .B(s01), .CI(1'b0),
.SUM(z0));
DW01_add #(16) add1 (
  .A(s10), .B(s11), .CI(1'b0),
.SUM(z1));
``
	``
	input [7:0] a, b;
	input [15:0] c0, c1;
	output [15:0] z0, z1;
	
	// single datapath with:
	// - automatic sharing of
	
	// multiplier
	// - implicit usage of carry-
	
	// save internally
	
	assign z0 = a * b + c0;
	assign z1 = a * b + c1;
	``

Complementing an Operand

- **Rule:** Do not complement (negate) operands manually by inverting all bits and adding a '1' (for example, "a_neg = ~a + 1"). Instead, arithmetically complement operands by using the '-' operator (for example, "a_neg = -a").
- **Rationale:** Simpler, higher level code. Manual complementing is not always recognized as an arithmetic operation and therefore can limit datapath extraction and result in *worse QoR*. Arithmetically complemented operands can easily be extracted as part of a bigger datapath.
- **Example:**

Bad QoR	Good QoR
| ```
input   signed   [7:0] a, b;
input   signed  [15:0] c;
input                  sign;
output signed [15:0] z;
wire    signed [15:0] p;

// manual complement prevents
// SOP extraction

assign p = a * b;

assign z = (sign ? ~p : p) +
   $signed({1'b0, sign}) + c;

// -> multiply + select + 3-
// operand add
``` | ```
input   signed   [7:0] a, b;
input   signed  [15:0] c;
input                  sign;
output signed [15:0] z;
wire    signed  [8:0] a_int;

// complement multiplier instead of
// product (cheaper)

assign a_int = sign ? -a : a;

assign z     = a_int * b + c;

// -> complement + SOP (multiply +
// add)
``` |

Appendix B
Examples

This chapter contains more complete versions of some of the sample code mentioned in the text.

State Machine Example

Simple Hierarchical State Machine in synthesizable SystemVerilog

This example shows how to code a hierarchical finite state machine in synthesizable SystemVerilog using tasks. Please see the notes following the code.

```
module foo (
  input bit clk, resetn, pkt_avail, pkt_out_fifo_full,
  input bit [31:0] data_in,
  output bit [15:0] data_out,
  output bit in_pkt_pop) ;

  struct packed {
    bit [7:0] destination ;
    bit [7:0] payload ;
  } input_packet ;

  enum { IDLE, GET_PKT, SEND_PKT} tx_state ;
  enum { GP_READ, GP_DONE } get_pkt_state ;
  enum { SP_DESTINATION, SP_PAYLOAD, SP_DONE } send_pkt_state;
```

M. Keating, *The Simple Art of SoC Design: Closing the Gap between RTL and ESL*, DOI 10.1007/978-1-4419-8586-6, © Synopsys, Inc. 2011

```verilog
// ------------ main state machine ----------------

always @ (posedge clk or negedge resetn) begin
   if (! resetn)begin
     in_pkt_pop <= 0;
     data_out <= 0;
     tx_state <= IDLE;
     get_pkt_state <= GP_READ;
     send_pkt_state <= SP_DESTINATION;
   end else begin
     case (tx_state)
       IDLE : if (pkt_avail) tx_state <= GET_PKT ;
       GET_PKT : begin
         get_pkt () ;
         if (get_pkt_state == GP_DONE) tx_state <= SEND_PKT ;
       end
         SEND_PKT : begin
           send_pkt () ;
         if (send_pkt_state == SP_DONE) tx_state <= IDLE ;
       end
     endcase
   end
end


// ---------- get_pkt sub state machine -------------

  task get_pkt() ;
    case (get_pkt_state)
      GP_READ : begin
        input_packet.destination <= data_in[31:16] ;
        input_packet.payload <= data_in[15:0] ;
        in_pkt_pop <= 1 ;
        get_pkt_state <= GP_DONE ;
      end
      GP_DONE : begin
        in_pkt_pop <= 0 ;
        get_pkt_state <= GP_READ ;
      end
    endcase
  endtask
```

```
// ---------- send_pkt sub state machine ------------

  task send_pkt() ;
    case (send_pkt_state)
      SP_DESTINATION : begin
        if ((!pkt_out_fifo_full)) begin
          data_out <= input_packet.destination ;
          send_pkt_state <= SP_PAYLOAD ;
        end
      end
      SP_PAYLOAD : begin
        if ((!pkt_out_fifo_full)) begin
          data_out <= input_packet.payload ;
          send_pkt_state <= SP_DONE ;
        end
      end
      SP_DONE : begin
        send_pkt_state <= SP_DESTINATION ;
      end
    endcase
  endtask

endmodule
```

Notes on the HSMF example:

1) There are many styles for writing hierarchical state machines in SystemVerilog, none of them right for every application. In general, there is a tradeoff between simplicity (ease of understanding) and optimality (minimum latency). This example was coded for simplicity.

2) The check for a sub state machine being done is of the form: **if** (get_pkt_ state == GP_DONE).This approach works only if we know that the state *GP_DONE* will take exactly one cycle. It is generally a good idea to keep the DONE state very simple as shown in this example.

3) The GP_DONE and SP_DONE cycles could be eliminated to save a cycle of latency, but at the cost of making the code more complex and prone to errors. The de-assertion of in_pkt_pop would need to be moved to the main state machine (GET_PKT section). And the testing to see if the sub state machines are done would become much more complex.

The goal of our proposed SystemVerilog state machine primitive is to resolve these problems and provide a structure that provides for simple code and minimum latency.

DCT Example Code:

C Version

The following is the original behavioral C code for the DCT:

```c
//(C) COPYRIGHT 2001 INSILICON CORPORATION
//ALL RIGHTS RESERVED

#include<stdlib.h>

int block[8][8];

int c[8][8]={
{23168, 32144, 30272, 27248, 23168, 18208, 12544, 6400},
{23168, 27248, 12544, 6400, 23168, 32144, 30272, 18208},
{23168, 18208, 12544, 32144, 23168, 6400, 30272, 27248},
{23168, 6400, 30272, 18208, 23168, 27248, 12544, 32144},
{23168, 6400, 30272, 18208, 23168, 27248, 12544, 32144},
{23168, 18208, 12544, 32144, 23168, 6400, 30272, 27248},
{23168, 27248, 12544, 6400, 23168, 32144, 30272, 18208},
{23168, 32144, 30272, 27248, 23168, 18208, 12544, 6400}
};

int s[8][8]={
{0, 0, 0, 0, 0, 0, 0, 0},
{0, 0, 0, 1, 1, 1, 1, 1},
{0, 0, 1, 1, 1, 0, 0, 0},
{0, 0, 1, 1, 0, 0, 1, 1},
{0, 1, 1, 0, 0, 1, 1, 0},
{0, 1, 1, 0, 1, 1, 0, 1},
{0, 1, 0, 0, 1, 0, 1, 0},
{0, 1, 0, 1, 0, 1, 0, 1}
};

int pixout;

void dct(){
   int y,x,u,v;
   int reg[8];

/* Horizontal */
```

```
for(y=0;y<8;y++){
    for(x=0;x<8;x++)
      reg[x]=0;

    for(x=0;x<8;x++){
      for(u=0;u<8;u++){
        v=block[y][x]*c[x][u];
        v+=2048;
        v>>=12;
        if(s[x][u]) v=-v;
        reg[u]+=v;
      }
    }

    for(x=0;x<8;x++) {
      block[y][x]=reg[x];
    }
  }


/* Vertical */
  for(y=0;y<8;y++){
    for(x=0;x<8;x++)
      reg[x]=0;

    for(x=0;x<8;x++)
      for(u=0;u<8;u++){
        v=block[x][y]*c[x][u];
        v+=131072;
        v>>=18;
        if(s[x][u])v=-v;
        reg[u]+=v;
      }

    for(x=0;x<8;x++){
      v=reg[x];
      v+=2;
      v>>=2;
      block[x][y]=v;
    }
  }
}
```

Final RTL Version

The following is the RTL Version of the DCT, condensed aggressively using SystemVerilog, including signed arithmetic, multi-dimensional arrays, functions and tasks:

```systemverilog
//-------------------------------------------------------
//
// (C) COPYRIGHT 2002-2009 SYNOPSYS, INC.
// ALL RIGHTS RESERVED
//
//-------------------------------------------------------
`timescale 1 ns / 1 ns // timescale for following modules

module dct (
   input clk,
   input start,
   input en, // enable
   input idct,
   input clr,
   input cend, // End of processing signal
   input signed [7:0] pixin, // Input pixel - DCT mode
   input signed [14:0] xv, // Input from DCTRam Vertical port
   output bit [14:0] yh, // Output to DCTRam Horizontal port
   output bit [10:0] zd, // pixel output - DCT mode
   output bit [5:0] memh, // DCTRam address - horizontal write
   output bit [5:0] memv, // DCTRam address - veritical read
   output bit ready); // Ready synchronization pulse

   bit [14:0] yv; // DCT vertical output before truncation
   bit [2:0] index;

//--Generate addresses for DCTRAM and increment index counter --
   bit fliph;
   bit flipv;

   task update_addr();
     if (start ) begin
       index <= 0;
       fliph <= 1;
       flipv <= 0;
       memh <= 6'b000111;
       memv <= 6'b000010;
     end else begin
```

```
      index <= index + 1;
      memh <= incr_addr (fliph, memh);
      memv <= incr_addr (flipv, memv);
      if (memh == 6'b111111) fliph <= ~fliph;
      if (memv == 6'b111111) flipv <= ~flipv;
    end
  endtask

  function automatic bit [5:0] incr_addr (bit flip,
    bit[5:0] addr); bit [5:0] addr_rev;

    if (!flip) incr_addr = addr + 1;
    else begin
      addr_rev = {addr[2:0], addr[5:3]};
      addr_rev++;
      incr_addr = {addr_rev[2:0], addr_rev[5:3]};
    end
  endfunction

//--------------- calculate sign --------------------
function automatic bit [7:0] get_sgn ();
  bit [7:0] sgn;
  sgn[0] = 0;
  case (index)
    3'b 000: sgn[7:1] = 7'b 0000000;
    3'b 001: sgn[7:1] = 7'b 1111100;
    3'b 010: sgn[7:1] = 7'b 0001110;
    3'b 011: sgn[7:1] = 7'b 1100110;
    3'b 100: sgn[7:1] = 7'b 0110011;
    3'b 101: sgn[7:1] = 7'b 1011011;
    3'b 110: sgn[7:1] = 7'b 0101001;
    3'b 111: sgn[7:1] = 7'b 1010101;
  endcase
  get_sgn = sgn;
endfunction

  //-------- declarations for pixel processing --------
  typedef bit [16:0] bit17;
  typedef bit [15:0] bit16;
  typedef bit [14:0] bit15;
  typedef bit [12:0] bit13;
  typedef bit [10:0] bit11;
```

```
// horizontal variables
bit15 [7:0] reg_x; // Accumulation registers
bit15 [7:1] sh; // Output shift registers
bit signed [10:0] x; // registered version of input pixel
// vertical variables
bit17 [7:0] reg_xv;
bit17 [7:1] shv ;
bit signed [14:0] xv_in;
bit en_p; // enable piped 1 cycle

//--------------- horizontal ---------------------
// Note: outputs are truncated to 13 bits
function bit13 [7:0] get_y (bit signed [10:0] x);
  bit signed [15:0] k7;
  bit signed [16:0] k6;
  bit signed [20:0] k5;
  bit signed [21:0] k3;
  bit signed [19:0] k2;
  bit signed [21:0] k1;
  bit signed [18:0] k0;

  k7 = x * 25 ;
  k6 = x * 49;
  k5 = k7 + (x * 544);
  k3 = x * 1703;
  k2 = k7 + (x * 448);
  k1 = k7 + (x * 1984);
  k0 = x * 181;

  get_y[0] = k0[18:6];
  get_y[1] = k1[21:9];
  get_y[2] = k2[19:7];
  get_y[3] = k3[21:9];
  get_y[5] = k5[20:8];
  get_y[6] = {k6[16], k6[16:5]};
  get_y[7] = {k7[15], k7[15], k7[15:5]};
  get_y[4] = 0;
endfunction
```

```
//----------- Map the vector for accumulation --------

function bit13 [7:0] get_crossh ( bit13 [7:0] y);
  case (index)
    3'd0: get_crossh[7:0] = {y[7],y[6],y[5],y[0],y[3],
                             y[2],y[1],y[0]};
    3'd1: get_crossh[7:0] = {y[5],y[2],y[1],y[0],y[7],
                             y[6],y[3],y[0]};
    3'd2: get_crossh[7:0] = {y[3],y[2],y[7],y[0],y[1],
                             y[6],y[5],y[0]};
    3'd3: get_crossh[7:0] = {y[1],y[6],y[3],y[0],y[5],
                             y[2],y[7],y[0]};
    3'd4: get_crossh[7:0] = {y[1],y[6],y[3],y[0],y[5],
                             y[2],y[7],y[0]};
    3'd5: get_crossh[7:0] = {y[3],y[2],y[7],y[0],y[1],
                             y[6],y[5],y[0]};
    3'd6: get_crossh[7:0] = {y[5],y[2],y[1],y[0],y[7],
                             y[6],y[3],y[0]};
    3'd7: get_crossh[7:0] = {y[7],y[6],y[5],y[0],y[3],
                             y[2],y[1],y[0]};
  endcase
endfunction



//---------------- Accumulate ----------------------
function bit15 [7:0] get_reg_x (bit signed [10:0] x);
  bit15 [7:0] tmp;
  bit13 bx[7:0];
  bit16 b[7:0];
  bit13 [7:0] sel;
  bit [7:0] sgn;

  sel = get_crossh (get_y(x));

  sgn = get_sgn();

  for (int i=0;i<8;i++) bx[i] = sgn[i] ? ~sel[i] : sel[i];
  for (int i=0;i<8;i++) b[i] = {bx[i][12],bx[i][12],
                               bx[i][12],bx[i]};

  if (index == 0) for (int i=0;i<8;i++) tmp[i] = b[i]
             [15:1] + b[i][0];
  else for (int i=0;i<8;i++) tmp[i] = reg_x[i] + b[i]
                [15:1] + b[i][0];
  get_reg_x = tmp;

endfunction
```

```
task update_sr();
  reg_x <= get_reg_x(x);
  if (index == 0) for (int i=1;i<8;i++) sh[i]<= reg_x[i];
  else for (int i=1;i<7;i++) sh[i]<= sh[i+1];
endtask

//----------------- vertical -----------------------
// Note: outputs are truncated to 16 bits
function bit16 [7:0] get_y_vert (bit signed [14:0] x);
  bit signed [19:0] k7;
  bit signed [20:0] k6;
  bit signed [24:0] k5;
  bit signed [25:0] k3;
  bit signed [23:0] k2;
  bit signed [25:0] k1;
  bit signed [22:0] k0;

  k7 = x * 25 ;
  k6 = x * 49;
  k5 = k7 + (x * 544);
  k3 = x * 1703;
  k2 = k7 + (x * 448);
  k1 = k7 + (x * 1984);
  k0 = x * 181;

  get_y_vert = 0;
  get_y_vert[0] = {k0[22], k0[22], k0[22], k0[22:10]};
  get_y_vert[1] = {k1[25], k1[25], k1[25], k1[25:13]};
  get_y_vert[2] = {k2[23], k2[23], k2[23], k2[23:11]};
  get_y_vert[3] = {k3[25], k3[25], k3[25], k3[25:13]};
  get_y_vert[5] = {k5[24], k5[24], k5[24], k5[24:12]};
  get_y_vert[6] = {k6[20], k6[20], k6[20],
                   k6[20],k6[20:9]};
  get_y_vert[7] = {k7[19], k7[19], k7[19], k7[19],
                   k7[19], k7[19:9]};
endfunction
```

```
/---------- Map the vector for accumulation ----------
function bit16 [7:0] get_crossv ( bit16 [7:0] y);
  case (index)
    3'd0: get_crossv[7:0] = {y[7],y[6],y[5],y[0],y[3],
                             y[2],y[1],y[0]};
    3'd1: get_crossv[7:0] = {y[5],y[2],y[1],y[0],y[7],
                             y[6],y[3],y[0]};
    3'd2: get_crossv[7:0] = {y[3],y[2],y[7],y[0],y[1],
                             y[6],y[5],y[0]};
    3'd3: get_crossv[7:0] = {y[1],y[6],y[3],y[0],y[5],
                             y[2],y[7],y[0]};
    3'd4: get_crossv[7:0] = {y[1],y[6],y[3],y[0],y[5],
                             y[2],y[7],y[0]};
    3'd5: get_crossv[7:0] = {y[3],y[2],y[7],y[0],y[1],
                             y[6],y[5],y[0]};
    3'd6: get_crossv[7:0] = {y[5],y[2],y[1],y[0],y[7],
                             y[6],y[3],y[0]};
    3'd7: get_crossv[7:0] = {y[7],y[6],y[5],y[0],y[3],
                             y[2],y[1],y[0]};
  endcase
endfunction

//------------- Accumulate and shift out -------------
function bit17 [7:0] get_reg_xv (bit signed [14:0] xv_in);
  bit17 [7:0] tmp;
  bit16 [7:0] bx;
  bit17 [7:0] b;
  bit16 [7:0] sel1;
  bit [7:0] sgn;

  sel1 = get_crossv (get_y_vert(xv_in));

  sgn = get_sgn();

  for (int i=0;i<8;i++) bx[i] = sgn[i] ? ~sel1[i] : sel1[i];
  for (int i=0;i<8;i++) b[i] = {bx[i][15], bx[i]};

  if (index == 0)for(int i=0;i<8;i++)
    tmp[i] = {b[i][16],b[i][16:1]} + b[i][0];
  else for (int i=0;i<8;i++)
    tmp[i] = reg_xv[i] +{b[i][16],b[i][16:1]} + b[i][0];

  get_reg_xv = tmp;

endfunction
```

```
task update_srv ();
  reg_xv <= get_reg_xv (xv_in);
  if (index == 0) for (int i=1;i<8;i++) shv[i]<= reg_xv[i];
  else for (int i=1;i<7;i++) shv[i] <= shv[i+1];
endtask

//-------------------- MAIN ------------------------
task main ();
  x <= {~pixin[7], (~pixin[7]), pixin[6:0], 2'b 00};
  en_p <= en;
  if (en) update_addr();
  if (en_p) update_sr();
  xv_in <= xv;
  if (en) update_srv();
endtask

always_comb
  if (index == 0) yh = reg_x[0];
  else yh = sh[1];

always_comb begin
  if (index == 0) yv = (reg_xv[0]+ 2) >>> 2;
  else yv = (shv[1]+ 2) >>> 2;
  zd = yv[10:0];
end

always_ff @(posedge clk) main();

endmodule
```

DCT Using Proposed SystemVerilog Extensions

This version is very close to the original C version, but becomes synthesizable with
the use of the proposed SystemVerilog extensions. Note that the constant-handling
optimizations from the RTL are not used, and that all variables are integers. This is
just to keep the code as simple as possible, and allow focus on the SystemVerilog
extensions. The constant-handling optimizations could certainly be added to this
code; it would just make the code a bit larger.

```
//----------------------------------------------------
//
// (C) COPYRIGHT 2002-2009 SYNOPSYS, INC.
// ALL RIGHTS RESERVED
//
//
```

```
//----------------------------------------------------------
module dct_top (
  input bit clk, resetn,
  input bit run,
  input bit mem_read_rdy,
  input int pixin,
  output int pixout);

  $clock posedge (clk);

  int c[8][8]='{
    '{23168, 32144, 30272, 27248, 23168, 18208, 12544, 6400 },
    '{23168, 27248, 12544, 6400, 23168, 32144, 30272, 18208},
    '{23168, 18208, 12544, 32144, 23168, 6400, 30272, 27248},
    '{23168, 6400, 30272, 18208, 23168, 27248, 12544, 32144},
    '{23168, 6400, 30272, 18208, 23168, 27248, 12544, 32144},
    '{23168, 18208, 12544, 32144, 23168, 6400, 30272, 27248},
    '{23168, 27248, 12544, 6400, 23168, 32144, 30272, 18208},
    '{23168, 32144, 30272, 27248, 23168, 18208, 12544, 6400 }
  };

  int s[8][8]='{
    '{0, 0, 0, 0, 0, 0, 0, 0},
    '{0, 0, 0, 1, 1, 1, 1, 1},
    '{0, 0, 1, 1, 1, 0, 0, 0},
    '{0, 0, 1, 1, 0, 0, 1, 1},
    '{0, 1, 1, 0, 0, 1, 1, 0},
    '{0, 1, 1, 0, 1, 1, 0, 1},
    '{0, 1, 0, 0, 1, 0, 1, 0},
    '{0, 1, 0, 1, 0, 1, 0, 1}
  };

int v_pixin;
  int tmp_x[8];
  int reg_x[8];
  int tmp_y[8];
  int reg_y[8];
  int mem[8][8];

//----------------- top (main) -----------------
state_machine dct();
  begin
    if (run) horizontal_dct();
  endstate_machine
```

```
//-------------- horizontal processing ---------------
state_machine horizontal_dct();
  done = 0;
  while (run) begin
    for (state_var y=0;y<8;y++) begin
      for(state_var x=0;x<8;x++) begin
        for(int u=0;u<8;u++) begin
          v=pixin*c[x][u];
          v+=2048;
          v>>>12;
          if(s[x][u]) v=-v;
          if (x ==0) reg_x[u]<=v;
          else if (x < 7) reg_x[u]<=reg_x[u]+v;
          else tmp_x[u] <= reg_x[u] + v;
        end
      end
      fork mem_write(y); join_none // can't fork again until
complete
    end
  end
  done = 1;
endstate_machine

//------------- write scratchpad memory ---------------
state_machine mem_write(int y);
  bit_ff flip;
  for(state_var x=0;x<8;x++) begin
    if (!flip) mem[y][x]<=tmp_x[x];
    else mem[x][y]<=tmp_x[x];
    if ((y == 7) && (x == 0) fork mem_read(); join_none
    if ((y == 7) && (x == 7) flip <= ~flip;
  end
endstate_machine


//-------------- read scratchpad memory --------------
state_machine mem_read();
  bit_ff flip = 0;
  bit_ff v_pixin_available = 0;
  for (state_var y=0;y<8;y++) begin
    for(state_var x=0;x<8;x++) begin
      if (!flip) v_pixin <= mem[y][x];
```

```
        else v_pixin <= mem[x][y];
        if ((y == 7) && (x == 7) flip <= ~flip;
        v_pixin_available <= mem_read_rdy;
        if ((y == 0) && (x == 0)) fork vertical_dct; join_none
      end
    end
endstate_machine

//--------------- vertical processing -----------------
state_machine vertical_dct();
  int v;
  for(state_var y=0;y<8;y++) begin
    for(state_var x=0;x<8;x++) begin
      wait (mem_read.v_pixin_available);
      for(int u=0;u<8;u++) begin
        v=v_pixin*c[x][u];
        v+=131072;
        v>>>18;
        if(s[x][u]) v=-v;
        reg_y[u]+=v;
        if (x == 0) reg_y[u]<=v;
        else if (x < 7) reg_y[u]<=reg_y[u]+v;
        else tmp_y[u] <= reg_y[u} + v;
      end
      fork output_pixels(); join_none
    end
  end
endstate_machine

//----------------- output results ------------------
state_machine output_pixels();
  for(state_var x=0;x<8;x++) begin
    v=tmp_y[x];
    v+=2;
    v>>>2;
    pixout <= v;
  end
endstate_machine

//----------------- invoke main state machine --------
dct();

endmodule
```

Appendix C
Preliminary Specification
for Extensions to SystemVerilog

This appendix includes preliminary specifications for some proposed extensions to the synthesizable subset of SystemVerilog. The first section addresses the issue of data types in SystemVerilog. The second section describes a hierarchical state machine construct.

Data Types in System Verilog

Overview

The most basic objects of digital design are:

- The flip-flop
- The combinational gate
- The latch
- The special signals clock and reset

None of these types are explicitly supported by SystemVerilog. The supported data types of bit, logic, reg and wire do not exactly map onto the design primitives listed above.

SystemVerilog introduced always_ff, always_comb, and always_latch to enable designers to specify exactly how assignments in these processes should be interpreted. Quoting from the LRM:

M. Keating, *The Simple Art of SoC Design: Closing the Gap between RTL and ESL*,
DOI 10.1007/978-1-4419-8586-6, © Synopsys, Inc. 2011

The SystemVerilog **always_ff** procedure can be used to model synthesizable sequential logic behavior. For example:

always_ff @ (**posedge** clock **iff** reset == 0 **or** **posedge** reset)
begin
r1 <= reset ? 0 : r2 + 1;
...
end

SystemVerilog provides a special **always_comb** procedure for modeling combinational logic behavior. For example:

always_comb
a = b & c;

SystemVerilog also provides a special **always_latch** procedure for modeling latched logic behavior. For example:

always_latch
if (ck) q <= d;

Limitation of SystemVerilog

These three new process types provide greatly improved simulation and synthesis semantics over Verilog. But they take the somewhat convoluted approach of defining a data type (a flop or combinational signal) by the process in which it is assigned a value.

A more straight-forward approach is to define SystemVerilog variables to be of the appropriate primitive type.

Proposal – New Module Type

We propose a special kind of module, called an *smodule*, which is exactly the same as a module in SystemVerilog except that the new data types are allowed in an smodule and only in an smodule. Because the new data types introduce some significant changes in assignment statements, a separate kind of module allows the user to specify exactly when and where these new data types and rules apply.

Having a separate kind of module also allows a stricter set of rules within an smodule, since we no longer need to allow mixing and matching syntaxes.

Proposal – New Data Types

We propose to use a similar syntax to define the following signal types:

bit_ff: A variable of type bit which can only be simulated/synthesized as an edge-triggered flop.

bit_comb: A variable of type bit which can only be simulated/synthesized as a combinational variable.

bit_latch: A variable of type bit which can only be simulated/synthesized as an level sensitive latch.

Similarly, we define **logic_ff**, **logic_comb**, and **logic_latch** as equivalent variables of type logic.

For types **bit_ff**, **bit_latch, logic_ff**, and **logic_latch** we allow initialization values to indicate the reset value of the variable. For example:

bit_ff [31:0] foo = 32'b0;

In addition, we provide for the specification of a default clock and a default reset to be used in a module.

$clock posedge clk specifies that the posedge of signal clk is the default clock for all flops and latches in this module.

$reset async negedge reset_n specifies that the signal reset_n is the default reset for all flops and latches in this module, that reset is asynchronous and occurs on the negedge of the signal reset_n.

$reset sync !reset_n specifies that the signal reset_n is the default reset for all flops and latches in this module, that reset is synchronous and occurs when the signal reset_n is low.

Usage

For the purpose of this section, we refer to bit_ff, bit_comb, bit_latch, logic_ff, logic_comb, logic_latch as extended bit/logic types.

The extended bit/logic types can be used only in an smodule. They can be used anywhere bit/logic can be used. That is, any object than can be declared to be of type bit (or logic) in a normal module can be declared to be an extended bit(or logic) type in an smodule. But the distinction between extended bit/logic types and normal bit/logic types becomes meaningful only when the variable appears on the left hand side of an assignment. That is, the compiler treats extended bit/logic data types exactly like normal bit/logic types except when they appear on the left hand side of an assignment operator.

Sequential and Combinational Processes

Explicit processes are not allowed in smodules. That is, no kind of *always* statement or *assign* statement is allowed. Since we have the extended data types, always_ff and always_comb are not required to disambiguate what an assignment means in terms of behavior and hardware structures that will be synthesized.

Operators

It will be an error if:

1. A variable of type bit_ff, bit_latch, logic_ff or logic_latch is on the left hand side of any assignment operator except the non-blocking assignment operator (<=).
2. A variable of type bit_comb or logic_comb is on the left hand side of any assignment operator except the blocking assignment operator (=).

Assignments Outside of Processes

In smodules, an assignment to an extended bit/logic type implies the appropriate process.
For example:

```
smodule foo (input bit [7:0] a,b );

    bit_comb [7:0] bar;

    bar = a && b;

endmodule
```

Is the equivalent of:

```
module foo (input bit [7:0] a,b );
bit [7:0] bar;

    always_comb bar = a && b;

endmodule
```

Sequential Assignments:

For implied sequential processes, it is necessary to know the clock, reset signal, and reset value. These are provided by $clock, $reset, and variable initialization.
For example:

```
smodule foo (
input bit clk, reset_n,
input bit [7:0] a,b );

$clock  posedge(clk); //all   flops   are   clocked
                // by this clock


$reset  sync  negedge resetn; //all  flops  are  reset
                // by this signal

bit_ff [7:0] bar = 8'h1f;

   bar <= a && b;

endmodule
```

Is the equivalent of:

```
module foo (
     input bit clk, reset_n,
     input bit [7:0] a,b );

bit [7:0] bar;

always_ff @ (posedge clk or negedge resetn) begin
     if (!resetn) begin
        bar <= 8'b1f;
     end else begin
        bar <= a && b;
     end
  end

endmodule
```

$clock

It will be an error if:

1. More than one $clock statement occurs in an smodule.
2. A $clock statement occurs after any assignment in the smodule. That is, the $clock is part of the declaration section at the beginning of the module, before any executable statement.
3. A $clock statement occurs in a task or function.

The $clock statement specifies the signal used for the clock and which edge is to be used:

$clock posedge|negedge signal_name

$reset

The same limitations described for $clock above apply also to the $reset statement.
The $reset statement consists of :

$reset sync|async posedge|negedge signal_name

For an implied sequential process, the reset value of the bit_ff or bit_latch is specified when the variable is declared:

```
bit_ff [7:0] foo = 8'hff;
```

If no reset value is specified in the declaration, the reset value of 0 will be used.

Semantics of Extended Data Types

Extended data types allow assignments to flip-flops and combinational logic to be mixed freely. For example:

```
if (cond) begin
    v1    = a || b;
    foo <= 8'b0;
    bar   = c + d;
end else begin
    bar   = c - d;
    foo <= 8'bff;
    v1    = a && b;
end
```

Since we can mix assignments, it makes no sense to view these as blocking or non-blocking assignments. In no sense does

```
v1    = a || b
```

block

```
foo <= 8'b0 or bar  = c + d.
```

Instead, we think of combinational assignments as occurring immediately, and sequential assignments occurring at the next clock. Thus the above code is equivalent to:

```
always_ff @ (posedge clk or negedge resetn) begin
  if (!resetn) begin
    foo <= (reset_value_of_foo);
  end else begin
    if (cond) begin
      foo <= 8'b0;
    end else begin
      foo <= 8'bff;
    end
  end
end

always_comb begin
  if (cond) begin
    v1   = a || b;
  end else begin
    v1   = a && b;
  end
end

always_comb begin
  if (cond) begin
    bar  = c + d;
  end else begin
    bar  = c - d;
  end
end
```

Or, in our more succinct extended data types:

```
if (cond) foo <= 8'b0;   else foo <= 8'bff;

if (cond) v1   = a || b; else v1   = a && b;

if (cond) bar  = c + d;  else begin bar   = c - d;
```

That is, objects of type bit_comb act like logic gates.

The classic example of blocking assignments is:

```
b = a;
a = c;
```

In a regular module, or in a function in an smodule, the result of this code is that b gets the old value of a, a's new value after it is assigned the value of c.

In an smodule, this code results in

```
b = c;
```

Which is what happens in real hardware.

The other classic case is:

```
a = 0;
c = 1;
b = a;
a = c;
```

In real hardware, and in smodules, this code makes no sense and is an error because it assigns "a" two different values.

In smodules, we allow one exception to this rule that combinational assignments always work exactly like real hardware. The exception is a default value within a scope (begin/end pair).

```
bit_comb [7:0] a;

if (condition1) begin
    a = 0;
    if (condition2)        a[0] = 1;
    else if (condition3) a[1] = 1;
end
```

In this case, the increased specificity of "a[0] = 1;" overrides the default "a = 0;" So if condition1 and condition2 are true,

```
a = 7'b00000001
```

Tasks

A standalone invocation of a task is considered an implied sequential process. So:

```
smodule foo (input bit clk, reset_n,);

$clock posedge(clk);
$reset sync negedge resetn;

task bar();
  a <= b + c;
endtask ;

bar ();

endmodule
```

Is the equivalent of:

```
module foo (input bit clk, reset_n);

task bar();
  a <= b + c;
endtask ;


always_ff @ (posedge clk) begin
  bar ();
end

endmodule
```

Functions

The behavior and rule governing functions are not changed.

Structure in SModules

One advantage of explicit processes like always_ff is that they allow a scope to be defined where local variables can be declared and related statements grouped together. This kind of mechanism for localizing information and hiding it from the rest of the code is important for writing good code.

For this reason, smodules allow (named or unnamed) begin/end blocks within the module.

Note: although most simulators support this feature, it is not part of the current SystemVerilog Language Reference Manual and is not supported by synthesis tools.

Example:

```
smodule foo (input bit clk, reset_n,
             input bit [7:0] in1,
             output bit_ff [7:0] out1);

$clock posedge(clk);
$reset sync negedge resetn;


begin: my_local_scope
  enum { id1, id2 } my_var;
  bit_ff current_id;

  if (input[0]) current_id <= id1;
  else current_id <= id2;

  out1 <= current_id + 3;
end

endmodule
```

Limitations:

The following limitations are a result of the definitions give above:

1. If an input to a smodule is declared to be an extended bit/logic type, no checking will be done on how that signal is driven. For example:

```
smodule foo (input bit_ff [7:0] a,b);
```

The fact that a and b are defined as bit_ff has no effect – we don't check that the module driving these inputs drives them from a flip-flop. (Note – this could be an interesting check in the future).

2. If a function is declared as type bit_ff, it is the same as declaring it as bit, since a function/task can never be the left hand side of an assignment.

Examples

Reference for Examples

The following (legal SystemVerilog) code serves as a reference for the following examples which use extended data types.

```
module foo (
  input bit            clk, resetn,
  input bit            data_valid,
  input bit [31:0]     data_in,
  output bit [35:0]    output_data ,
  output bit           hdr_dword_push
  );


  always_ff @ (posedge clk or negedge resetn) begin
    if (!resetn) begin
      output_data <= 36'h0;
    end else begin
      output_data <= data_in;
    end
  end

  always_comb hdr_dword_push = data_valid;

endmodule
```

Example 1 – Extended data types in processes

```
smodule foo (
  input bit              clk, resetn,
  input bit              data_valid,
  input bit [31:0]       data_in,
  output bit_ff [35:0]   output_data = 36'h0,
  output bit_comb        hdr_dword_push
  );

  $clock posedge(clk);

  $reset async negedge(resetn);

  always_ff output_data <= data_in; // simpler

  always_comb hdr_dword_push = data_valid;
endmodule
```

Example 2 – Extended data types outside of processes

The **always_ff** and **always_comb** are unnecessary in the previous example,
because the signal declarations convey the same information.

```
smodule foo (
  input bit              clk, resetn,
  input bit              data_valid,
  input bit [31:0]       data_in,
  output bit_ff [35:0]   output_data = 36'h0,
  output bit_comb        hdr_dword_push
  );

  $clock posedge(clk);
  $reset async negedge(resetn);

  output_data <= data_in;

  hdr_dword_push = data_valid;
endmodule
```

Example 3 - A slightly more complex example

```
smodule foo (
  input bit              clk, resetn,
  input bit              data_valid,
  input bit [31:0]       data_in,
  output bit_ff [35:0] output_data = 36'h0,
  output bit_comb        hdr_dword_push
  );

  $clock posedge(clk);
  $reset async negedge(resetn);

  if (data_valid) output_data <= data_in;
  else for (int i=0; i<8; i++) output_data <= i;

  hdr_dword_push = data_valid;
endmodule
```

Hierarchical State Machine Primitive

Overview

The state machine is the most important element in control code. Today engineers code state machines in many different styles. This limits the optimizations synthesis can do and the analysis that verification tools can do. It also leads to state machines that are more complex and difficult for humans to understand. Although about 5-10% of RTL designers have used hierarchical state machines, the difficulty in coding them limits the adoption of this very effective tool for structuring control code.

The proposed hierarchical state machine (HSM) mechanism is similar to a function or a task, in that it is a scope. This allows local variables and types to be defined. Like functions and tasks, state machines can call other state machines, leading to hierarchical state machines.

Fundamental Characteristics of a State Machine

1. A state machine is a fully static object. State is persistent and (can be declared to be) local to the state machine.
2. State is a primitive design concept, so the state machine is not a function or task, although it can be coded using functions and tasks.

3. State is a property of a module – a module is in exactly one state at any time during operation.
4. State transitions are synchronous events. All state transitions in a given state machine occur on the same edge (pos or neg) of the same clock.
5. In a given state, both synchronous and combinational outputs are driven to new values. The synchronous events occur on the same edge (pos or neg) of the same clock as the state transitions.
6. The state machine primitive must support hierarchical state machines.

Translation Example – Simple State Machine

The features of the state machine primitive are described in Chapter 11. Here we give an example of a very simple state machine and how it might be translated into legal SystemVerilog. The following state machine:

```
state_machine  tctrl();
  bit flag1;
  state_type { TX_IDLE, TX_MPKT, TX_SKP } state;

  begin
    flag1 = 0;
    case (state)
      TX_IDLE: begin
        output_data <= 0;
        state <= TX_SKP;
      end
      TX_SKP: begin
        flag1 = 1;
        skp_sm();
        if (skp_sm.done) state <= TX_IDLE;
      end
      default: ;
    endcase
  end
endstate_machine
```

(continued)

(continued)

```
state_machine skp_sm
  enum { IDLE, NEXT, DONE } state;
  bit done ;
  bit flag2;

  begin
    done = 0;
    flag2 = 0;
    case (state)
      IDLE: state <= NEXT;
      NEXT: begin
        flag2 = 1;
        state <= DONE;
      end
      DONE: begin
        done = 1;
        state <= IDLE;
      end
    endcase
  end

endstate_machine
```

Gets translated to:

```
  bit tctrl_flag1;
  enum { TX_IDLE, TX_MPKT, TX_SKP } tctrl_state;

  task tctrl(); //body of sm
    case (tctrl_state)
      TX_IDLE: begin
        output_data <= 0;
        tctrl_state <= TX_SKP;
      end
      TX_SKP: begin
        skp_sm.main();
        if (skp_sm.done) tctrl_state <= TX_IDLE;
      end
      default: ;
    endcase
  endtask
```

(continued)

(continued)

```
always_comb if (tctrl_state == TX_SKP)
  tctrl_flag1 = 1;
else
  tctrl_flag1 = 0;


always @(posedge clk) tctrl();


enum {SKP_IDLE,SKP_NEXT,SKP_DONE} skp_sm_state;
bit skp_done ;
bit skp_flag2;

task skp_sm();
  case (skp_sm_state)
    SKP_IDLE: skp_sm_state <= SKP_NEXT;
    SKP_NEXT: skp_sm_state <= SKP_DONE;
    SKP_DONE: skp_sm_state <= SKP_IDLE;
  endcase
endtask

always_comb if (skp_sm_state == SKP_NEXT)
 skp_flag2 = 1;
else
 skp_flag2 = 0;

always_comb if (skp_sm_state == SKP_DONE)
 skp_done = 1;
else
 skp_done = 0;
```

Appendix D
More on High Level SystemVerilog

This chapter describes some additional high-level constructs in SystemVerilog and their potential role in raising the level of abstraction in design.

Interfaces

The interface construct in SystemVerilog can be a useful tool in creating more structured code. The interface construct is described in detail in [15] and [16].

The idea behind interfaces is that we can go beyond grouping interface signals together – which is easily and efficiently done using structs. With interfaces, we can move some functionality, especially functionality that is communication-dependent, from the module into an interface.

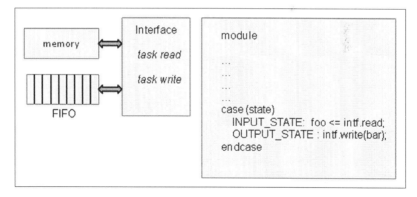

Figure D-1   Using the interface construct to encapsulate the memory and FIFO read/write tasks.

M. Keating, *The Simple Art of SoC Design: Closing the Gap between RTL and ESL*,
DOI 10.1007/978-1-4419-8586-6, © Synopsys, Inc. 2011

Figure D-1 illustrates this idea. Imagine we are in the early stages of a design, and we don't yet know if we are going to interface to a memory or a FIFO. We can use a SystemVerilog interface to encapsulate the read and write functions. We can then code the module using generalized read and write functions, which call functions (tasks) in the interface. Now we can change our minds about whether to use a FIFO or memory, and we need only change the code in the interface. The module itself remains unchanged.

This approach has proven very useful in writing test benches and behavioral code. With the current synthesizable subset, however, it is not possible to put sequential code in the interface block. This means that if the read or write takes multiple clock cycles, it is not really possible to isolate this behavior in the interface. This is another case where, hopefully, the synthesizable subset of SystemVerilog will be extended.

Parameters of Type Type

Another interesting high-level capability of SystemVerilog is the parameter of type type. Here we can specify the type of an object via a parameter. As in the case of the interface, this construct allows us to defer decisions. For instance, if we are not sure whether an interface is going to be a DMA interface or a register read-write interface, we can use parameters of type *type* to defer the decision but carry on with writing the rest of the code.

Example D-1 shows a (very simple) example of using parameters of type type. Example D1: Parameters of type type are synthesizable in the Synopsys synthesis tools.

```
typedef enum bit {READ, WRITE} rw_type;

typedef struct packed{
    bit [14:0] target_addr;
    bit [7:0] length;
} dma_inst_type ;

typedef struct packed {
    rw_type read_write;
    bit [14:0] addr;
    bit [15:0] data;
  } rw_instr_type;

interface automatic instruction_intf
    #(parameter type fifo_type = dma_inst_type)
    (input clk);
```

(continued)

```
    bit fifo_empty;
    bit fifo_pop;
    bit instr_avail;
    fifo_type instruction;

    modport fifo (input fifo_pop, clk,
                  output fifo_empty, instruction);

    modport master(input instruction, instr_avail,
                         clk,
                     import task get_instruction());

    task get_instruction ();
        if (!fifo_empty && !fifo_pop) begin
            fifo_pop <= 1'b1;
            instr_avail <= 1'b1;
        end else begin
            fifo_pop <= 1'b0;
            instr_avail <= 1'b0;
        end
    endtask
endinterface : instruction_intf

module main
 #(parameter type instruction_type = dma_inst_type)
  (instruction_intf.master di);

  instruction_type instruction;

  always @(posedge di.clk) begin
    di.get_instruction();
    if (di.instr_avail)
      instruction <= di.instruction;
    end
endmodule
```

Example D-1 (continued)

FIFOs

The rest of this appendix explores using the queue construct in SystemVerilog to create a synthesizable fifo primitive for SystemVerilog.

Queue Overview

The proposed FIFO primitive is built on the (fixed size) queue feature of SystemVerilog. The queue is not currently synthesizable. The goal of this project is define a subset/extension of the queue that provides a synthesizable FIFO primitive. In SystemVerilog, the queue in general is defined by:

byte q1[$]; // A queue of bytes

A fifo requires a fixed size queue, which is defined by:

bit q2[$:255]; // A queue whose maximum size is 256 bits

The queue in SV allows operators to operate directly on queues. The queue also comes with the following methods.

size()

The size() method returns the number of items in the queue. If the queue is empty, it returns 0.

insert()

The insert() method inserts the given item at the specified index position. For instance, Q.insert(i, e) is equivalent to: Q = {Q[0:i-1], e, Q[i,$]}

delete()

The delete() method deletes the item at the specified index position.

pop_front()

The pop_front() method removes and returns the first element of the queue.

pop_back()

The pop_back() method removes and returns the last element of the queue.

push_front()

The push_front() method inserts the given element at the front of the queue.

push_back()

The push_back() method inserts the given element at the end of the queue.

FIFO Definition

The synthesizable FIFO is:

* A fixed size queue
* No direct operations are allowed on the queue. The data in a fifo can only be accessed through its methods.

The supported queue methods are:

* size()
* pop_back()
* push_front()

In addition, the following new methods are provided:

* push (equivalent to push_front)
* pop (equivalent to pop_back)
* empty (equivalent to size() == 0)
* full (equivalent to size() == depth of fifo)
* move_write_pointer(count): moves the write pointer by count (so it moves the pointer forward if count is positive, moves it back if it is negative)
* move_read_pointer(count): moves the read pointer by count (so it moves the pointer forward if count is positive, moves it back if it is negative)

FIFO Examples

```
module fifo_test (
  input bit clk, reset_n,
  input data_valid,
  input byte inbyte);

  byte fifo [$:8];  // fifo is the variable name
                    // not a reserved word
  byte outbyte[3:0];

always_ff @(posedge clk) begin
```

(continued)

```
    if (data_valid && (fifo.size < 8))
      fifo.push_front(inbyte);
  end

  always_ff @(posedge clk) begin
    if (fifo.size > 4)
      for (int i = 0; i < 4; i++)
        outbyte[i] <= fifo.pop_back();
  end
endmodule
```

Example D-2 (continued)

Example D-2 shows the use of a bounded queue as a fifo. This is completely legal
SystemVerilog code – it simulates correctly. But it is not currently
synthesizable.
Example D-3 shows how to use a queue as a fifo in the case where it is read and
written by two different modules.

```
module fifo_test (
  input bit clk, reset_n,
  input data_valid,
  input byte inbyte);

    byte fifo [$:8];

  fifo_writer U_fifo_writer(.*);
  fifo_reader U_fifo_reader(.*);
endmodule

module fifo_writer (
  input bit clk, reset_n,
  input data_valid,
  input byte inbyte);

  always_ff @(posedge clk) begin
    if (data_valid &&(fifo_test.fifo.size < 8))
      fifo_test.fifo.push_front(inbyte);
```

(continued)

```
    end
 endmodule

module fifo_reader (
  input bit clk, reset_n,
  input data_valid,
  input byte inbyte);

  byte outbyte[3:0];

  always_ff @(posedge clk) begin
    if (fifo_test.fifo.size > 4)
      for (int i = 0; i < 4; i++)
        outbyte[i] <= fifo_test.fifo.pop_back();
  end
 endmodule
```

Example D-3 (continued)

Example D-4 shows a queue used as a fifo with the fifo in the interface.

```
interface fifo_interface ();
  byte fifo [$:8];

  modport master (import full, push);
  modport slave  (import empty, half_full, pop);

  function empty ();
    empty = (fifo.size() == 0);
  endfunction

  function full ();
    full = (fifo.size() == 8);
  endfunction

  task push (byte inbyte);
    fifo.push_front(inbyte);
  endtask

  function byte pop ();
    pop = fifo.pop_back();
  endfunction
```

(continued)

```systemverilog
  function half_full ();
     half_full = (fifo.size() >= 4);
  endfunction
endinterface : fifo_interface

module fifo_writer (
  input bit clk, reset_n,
  input data_valid,
  input byte inbyte,
  fifo_interface.master fifo);

  always_ff @(posedge clk)
     if (data_valid && !fifo.full())fifo.push(inbyte);

endmodule


module fifo_reader (
  input bit clk, reset_n,
  input data_valid,
  input byte inbyte,
  fifo_interface.slave fifo);

  byte outbyte;

  always_ff @(posedge clk)
     if (fifo.half_full())outbyte <= fifo.pop();

endmodule

module fifo_test (
  input bit clk, reset_n,
  input data_valid,
  input byte inbyte);

  fifo_interface fifo();

  fifo_writer U_fifo_writer(.*);
  fifo_reader U_fifo_reader(.*);

endmodule
```

Example D-4 (continued)

References

[1] Miller, G. A. (1956). "The magical number seven, plus or minus two: Some limits on our capacity for processing information". *Psychological Review* 63 (2): 81–97.

[2] Lorenz, Edward N. "Deterministic Nonperiodic Flow". *Journal of the Atmospheric Sciences* **20** (2): 130–141 (March 1963). Note: In spite of the title listed above, the effect has long been referred to as the butterfly affect. According to Lorenz, when he failed to provide a title for a talk he was to present at the 139th meeting of the American Association for the Advancement of Science in 1972, Philip Merilees concocted *Does the flap of a butterfly's wings in Brazil set off a tornado in Texas?* as the title.

[3a] Miller, J. and Page, S. Complex Adaptive Systems: An Introduction to Computational Models of Social Life (Princeton Studies in Complexity) Princeton University Press March 5, 2007

[3b] Bar_Yam, Y., Minai, A. Unifying Themes in Complex Systems: Proceedings of the Second International Conference on Complex Systems

[4a] http://www.sei.cmu.edu/library/abstracts/news-at-sei/wattsnew20043.cfm

[4b] Humphrey, W. A Discipline for Software Engineering. Addison Wesley, 1995.

[4c] http://weblogs.java.net/blog/2006/03/10/economics-quality

[5] McConnell, S. Code Complete: A Practical Handbook of Software Construction Microsoft Press; 2nd edition (June 9, 2004)

[6a] http://www.eetimes.com/design/automotive-design/4004785/Leveraging-system-models-for-RTL-functional-verification

[6b] http://ieeexplore.ieee.org/stamp/stamp.jsp?arnumber=01200580

[7] Jones,C. Applied Software Measurement: Global Analysis of Productivity and Quality by Capers Jones McGraw-Hill Osborne Media; 3 edition (April 11, 2008)

[8] Sutherland, S. et al. SystemVerilog for Design Second Edition: A Guide to Using SystemVerilog for Hardware Design and Modeling [Hardcover]Springer; 2nd edition (July 20, 2006)

[9] Keating, M. and Bricaud, P. Reuse Methodology Manual for System-on-a-Chip Designs Springer; 3rd edition (September 11, 2007)

[10] Harel, D. and Politi, M. Modeling Reactive Systems With Statecharts : The Statemate Approach McGraw-Hill Companies (October 8, 1998)

[11] Samek, M. Practical Statecharts in C/C++ CMPBooks 2008

[12] Hennessy, J. and Patterson, D. Computer Architecture: A Quantitative Approach, 3rd Edition Morgan Kaufmann (May 31, 2002)

[13] Fowler, M. and Parsons, R. Domain-Specific Languages **Martin Fowler** (Author) Addison-Wesley Professional; (October 4, 2010)

[14] Dijkstra, E. "Go To Statement Considered Harmful" (PDF). *Communications of the ACM* **11** (3): 147–148 (March 1968).

[15] Jones, C. Applied Software Measurement: Global Analysis of Productivity and Quality McGraw-Hill Osborne Media; 3 edition (April 11, 2008)

[16a] A. Koelbl, R. Jacoby, H. Jain, C. Pixley, "Solver Technology for System-level to RTL Equivalence Checking", Proceedings of DATE 2009, Munich, Germany

[16b] A. Koelbl, J.Burch, C. Pixley, "Memory Modeling in ESL-RTL Equivalence Checking, Proceedings of DAC 2007, San Diego, CA.

[16c] Alfred Koelbl, C.Pixley, "Constructing Efficient Formal Models from High-Level Descriptions Using Symbolic Simulation", Intl. J. of Parallel Programming, Volume 33, Number 6, December 2005, pp. 645–666

[17] McCabe, T. "A Complexity Measure" IEEE Transactions on Software Engineering Vol. 2, No. 4, p. 308 (1976) (http://www.literateprogramming.com/mccabe.pdf)

[18] Sutherland, S. et al., SystemVerilog for Design, Second Edition. Springer, 2006.

[19] Spear, C. SystemVerilog for Verification, Second Edition. Springer, 2008.

[20] http://www.arm.com/products/system-ip/amba/amba-open-specifications.php

[21] http://www.ocpip.org/home.php

[22] Philippe Coussy, Daniel D. Gajski, Michael Meredith, Andres Takach, "An Introduction to High-Level Synthesis," IEEE Design and Test, July/August 2009

[23] Marı´a Carmen Molina, Rafael Ruiz-Sautua, Alberto Del Barrio, and Jose´ Manuel Mendı´as, "Subword Switching Activity Minimization to Optimize Dynamic Power pConsumption", IEEE Design and Test, July/August 2009

[24] http://www.ghs.com/products/doublecheck.html

[25] http://csse.usc.edu/csse/research/COCOMOII/cocomo_main.html

[26] Boehm, B. Software Engineering Economics, Prentice Hall (November 1, 1981)

[27] Jones, C. Estimating Software Costs: Bringing Realism to Estimating, McGraw-Hill Osborne Media; 2 edition (April 19, 2007)

[28] https://solvnet.synopsys.com/retrieve/015771.html

Index